T0292002

Advanced mathematical methods for engineering and science students

Advanced mathematical methods for engineering and science students

G. STEPHENSON

*Emeritus Reader in Mathematics,
Imperial College, London*

and

P. M. RADMORE

*Lecturer, Department of Electronic and
Electrical Engineering,
University College London*

CAMBRIDGE
UNIVERSITY PRESS

Published by the Press Syndicate of the University of Cambridge
The Pitt Building, Trumpington Street, Cambridge CB2 1RP
40 West 20th Street, New York, NY 10011–4211, USA
10 Stamford Road, Oakleigh, Melbourne 3166, Australia

First published 1990
Reprinted 1993

British Library Cataloguing in Publication Data

Stephenson, G. (Geoffrey), *1927–*
Advanced mathematical methods for engineering and science students.
1. Mathematics
I. Title II. Radmore, P. M.
510

Library of Congress Cataloguing in Publication Data

Stephenson, G. (Geoffrey), 1927–
Advanced mathematical methods for engineering and science students / G. Stephenson and P. M. Radmore
p. cm.
ISBN 0-521-36312-8. — ISBN 0-521-36860-X (pbk)
1. Mathematics. I. Radmore, P. M. II. Title
QA39.2.S745 1990
510—dc20 89-36870 CIP

ISBN 0 521 36312 8 hardback
ISBN 0 521 36860 X paperback

Transferred to digital printing 2002

Contents

Preface

This book contains a selection of advanced topics suitable for final year undergraduates in science and engineering, and is based on courses of lectures given by one of us (G. S.) to various groups of third year engineering and science students at Imperial College over the past 15 years. It is assumed that the student has a good understanding of basic ancillary mathematics. The emphasis in the text is principally on the analytical understanding of the topics which is a vital prerequisite to any subsequent numerical and computational work. In no sense does the book pretend to be a comprehensive or highly rigorous account, but rather attempts to provide an accessible working knowledge of some of the current important analytical tools required in modern physics and engineering. The text may also provide a useful revision and reference guide for postgraduates.

Each chapter concludes with a selection of problems to be worked, some of which have been taken from Imperial College examination papers over the last ten years. Answers are given at the end of the book.

We wish to thank Dr Tony Dowson and Dr Noel Baker for reading the manuscript and making a number of helpful suggestions.

G.S.
Imperial College, London
P.M.R.
University College London

Notes to the reader

1. The symbol ln denotes logarithm to base e.
2. The end of a worked example is denoted by ◢

1
Suffix notation and tensor algebra

1.1 Summation convention

We consider a rectangular cartesian coordinate system with unit vectors $\mathbf{i}$, $\mathbf{j}$ and $\mathbf{k}$ along the three coordinate axes x, y and z respectively. For convenience, we relabel these unit vectors $\mathbf{e}_1$, $\mathbf{e}_2$ and $\mathbf{e}_3$ and denote the coordinate axes by x_1, x_2 and x_3 (Figure 1.1). A typical vector $\mathbf{a}$ with components a_1, a_2 and a_3 in cartesian coordinates can then be written as

$$\mathbf{a} = \sum_{i=1}^{3} a_i \mathbf{e}_i. \tag{1.1}$$

Instead of writing the summation sign in (1.1) every time we have an expression of this kind, we can adopt the summation convention: whenever an index occurs precisely twice in a term, it is understood that the index is to be summed over its full range of possible values without the need for explicitly writing the summation sign Σ. Hence (1.1), with this convention, is

$$\mathbf{a} = a_i \mathbf{e}_i, \tag{1.2}$$

where summation over i is implied ($i = 1, 2, 3$). Since the components a_i of $\mathbf{a}$ are given by the dot-product of $\mathbf{a}$ with each of the unit vectors $\mathbf{e}_i$, then $a_i = \mathbf{a} \cdot \mathbf{e}_i$ ($i = 1, 2, 3$) and (1.2) can be written

$$\mathbf{a} = (\mathbf{a} \cdot \mathbf{e}_i)\mathbf{e}_i, \tag{1.3}$$

again adopting the summation convention.

Example 1 If vectors **c** and **d** have components c_i and d_i ($i = 1, 2, 3$) respectively, then

$$\mathbf{c}\,.\,\mathbf{d} = \sum_{i=1}^{3} c_i d_i = c_i d_i. \quad ◢ \tag{1.4}$$

Example 2

$$\sum_{s=1}^{2} a_s x_s = a_1 x_1 + a_2 x_2 = a_i x_i. \quad ◢ \tag{1.5}$$

Example 3 Consider the term $a_{ij} b_i b_j$ in which i and j both occur twice. The summation convention implies summation over i and j independently. Hence, if i and j run from 1 to 2,

$$a_{ij} b_i b_j = a_{11} b_1 b_1 + a_{12} b_1 b_2 + a_{21} b_2 b_1 + a_{22} b_2 b_2 \tag{1.6}$$

$$= a_{11} b_1^2 + a_{22} b_2^2 + b_1 b_2 (a_{12} + a_{21}). \quad ◢ \tag{1.7}$$

The usual rules apply when the summation convention is being used, that is,

$$a_i b_i = b_i a_i \tag{1.8}$$

and

$$a_i(b_i + c_i) = a_i b_i + a_i c_i. \tag{1.9}$$

We see in (1.9) that although i occurs three times in the left-hand side it only occurs twice in each *term* and therefore summation over i is implied. Also

$$(a_i b_i)(c_j d_j) = (a_i c_j)(b_i d_j), \tag{1.10}$$

Figure 1.1

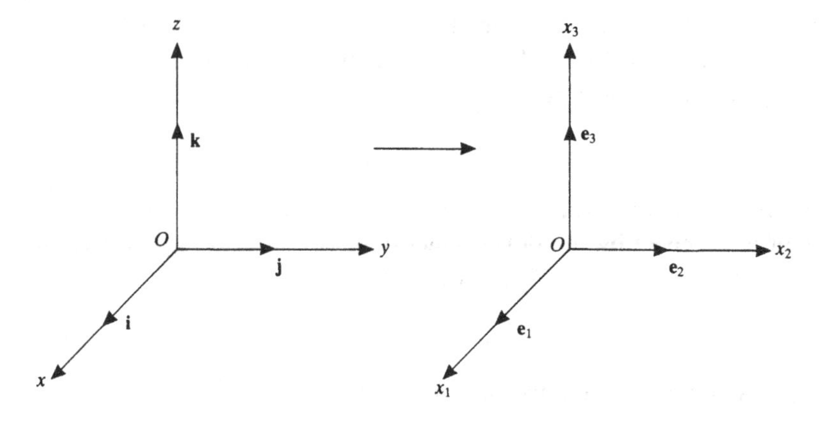

that is, the order of the factors is unimportant and summation is implied over both i and j. It would be wrong, however, to write (1.10) as $a_i b_i c_i d_i$ since the index occurs four times and therefore no summation over i would be implied.

Example 4

$$\left(\sum_{s=1}^{2} a_s x_s\right)^2 = (a_1 x_1 + a_2 x_2)^2 \tag{1.11}$$

$$= a_1^2 x_1^2 + 2a_1 a_2 x_1 x_2 + a_2^2 x_2^2. \tag{1.12}$$

Using the summation convention we can write this as

$$\left(\sum_{i=1}^{2} a_i x_i\right)\left(\sum_{j=1}^{2} a_j x_j\right) = a_i x_i a_j x_j, \tag{1.13}$$

where both i and j appear twice and consequently are both summed over the values 1 and 2. ◢

1.2 Free and dummy indices

Consider the following set of n linear equations for the quantities $x_1, x_2, \ldots, x_n$ with (constant) coefficients $a_{11}, a_{12}, \ldots, a_{1n}, a_{2n}, \ldots, a_{nn}$:

$$\left.\begin{aligned} a_{11}x_1 + a_{12}x_2 + \ldots + a_{1n}x_n &= c_1, \\ a_{21}x_1 + a_{22}x_2 + \ldots + a_{2n}x_n &= c_2, \\ \vdots \qquad \vdots \qquad\qquad \vdots \quad &\quad \vdots \\ a_{n1}x_1 + a_{n2}x_2 + \ldots + a_{nn}x_n &= c_n, \end{aligned}\right\} \tag{1.14}$$

where $c_1, c_2, \ldots, c_n$ are given constants. This set of equations can be written as

$$\left.\begin{aligned} \sum_{j=1}^{n} a_{1j}x_j &= c_1, \\ \sum_{j=1}^{n} a_{2j}x_j &= c_2, \\ \vdots \qquad &\quad \vdots \\ \sum_{j=1}^{n} a_{nj}x_j &= c_n. \end{aligned}\right\} \tag{1.15}$$

By introducing the index i, (1.15) can be expressed in the more compact form

$$\sum_{j=1}^{n} a_{ij}x_j = c_i, \quad (i = 1, 2, \ldots, n). \tag{1.16}$$

Using the summation convention, we can write (1.16) as

$$a_{ij}x_j = c_i. \tag{1.17}$$

Indices which occur twice, so that summation over them is implied (j in (1.17)), are called dummy indices, while indices which can have any value (i in (1.17)) are called free indices. We note that (1.17) could equally have been written as $a_{pq}x_q = c_p$, where summation over the dummy index q is implied and p is free.

Example 5 Suppose we are given constants a_{ij} ($i, j = 1, 2, 3$) and the function $\phi = a_{ij}x_ix_j$ (where summation over both i and j is implied). We wish to calculate the quantities $\partial\phi/\partial x_s$, where s is a free index (equal to 1, 2 or 3). Then by the chain rule for differentiation

$$\frac{\partial\phi}{\partial x_s} = a_{ij}\frac{\partial x_i}{\partial x_s}x_j + a_{ij}x_i\frac{\partial x_j}{\partial x_s}. \tag{1.18}$$

Since the x_i are independent variables, $\partial x_i/\partial x_s$ is 1 if $i = s$ and zero otherwise. Hence in the first term the i-summation has only one non-zero term (when $i = s$). Similarly the j-summation in the second term has only one non-zero term (when $j = s$). Thus

$$\partial\phi/\partial x_s = a_{sj}x_j + a_{is}x_i, \tag{1.19}$$

where j is summed in the first term and i in the second. These dummy indices can be given any letter we choose so that, replacing the dummy index j by i in the first term of (1.19),

$$\frac{\partial\phi}{\partial x_s} = a_{si}x_i + a_{is}x_i = (a_{si} + a_{is})x_i, \tag{1.20}$$

where i is now the dummy index and s is the free index. If a_{is} is symmetric so that $a_{is} = a_{si}$ then

$$\partial\phi/\partial x_s = 2a_{is}x_i, \tag{1.21}$$

whereas if a_{is} is skew- (or anti-) symmetric so that $a_{is} = -a_{si}$ then

$$\phi = 0 \quad \text{and} \quad \partial\phi/\partial x_s = 0. \; ◢ \tag{1.22}$$

1.3 Special symbols

1. Kronecker delta

The Kronecker delta symbol δ_{ij} is defined by

$$\delta_{ij} = \mathbf{e}_i \cdot \mathbf{e}_j = \begin{cases} 1 & \text{if} \quad i = j, \\ 0 & \text{if} \quad i \neq j. \end{cases} \tag{1.23}$$

We note that $\delta_{ij}\mathbf{e}_j = \delta_{i1}\mathbf{e}_1 + \delta_{i2}\mathbf{e}_2 + \delta_{i3}\mathbf{e}_3$ and that only one of these three terms is non-zero depending on i; for example $\delta_{11} = 1$, $\delta_{12} = \delta_{13} = 0$. Hence $\delta_{ij}\mathbf{e}_j = \mathbf{e}_i$. Also, as in (1.3),

$$\mathbf{a} = (\mathbf{a} \cdot \mathbf{e}_i)\mathbf{e}_i = (\mathbf{a} \cdot \mathbf{e}_j\delta_{ij})(\mathbf{e}_k\delta_{ik}). \tag{1.24}$$

Consider the quantity $\delta_{ij}\delta_{jk}$. Then

$$\delta_{ij}\delta_{jk} = (\mathbf{e}_i \cdot \mathbf{e}_j)(\mathbf{e}_j \cdot \mathbf{e}_k) \tag{1.25}$$

$$= (\mathbf{e}_i \cdot \mathbf{e}_1)(\mathbf{e}_1 \cdot \mathbf{e}_k) + (\mathbf{e}_i \cdot \mathbf{e}_2)(\mathbf{e}_2 \cdot \mathbf{e}_k) + (\mathbf{e}_i \cdot \mathbf{e}_3)(\mathbf{e}_3 \cdot \mathbf{e}_k). \tag{1.26}$$

Now if $i \neq k$, then at most one of the two brackets in each term can be non-zero and hence each term is zero. If $i = k$ then one term has both brackets non-zero and equal to 1. Hence $\delta_{ij}\delta_{jk}$ is zero if $i \neq k$ and is 1 if $i = k$. This is just the definition of δ_{ik} (see (1.23)) and so

$$\delta_{ij}\delta_{jk} = \delta_{ik}. \tag{1.27}$$

Further consider the expression $\delta_{rs}A_{pqs}$, where r, p and q are free indices. Then

$$\delta_{rs}A_{pqs} = \delta_{r1}A_{pq1} + \delta_{r2}A_{pq2} + \delta_{r3}A_{pq3}. \tag{1.28}$$

Only one of these terms is non-zero depending on the value of r. Hence

$$\delta_{rs}A_{pqs} = A_{pqr}. \tag{1.29}$$

2. *The alternating symbol*

The alternating symbol ϵ_{ijk} is defined as

$$\epsilon_{ijk} = \mathbf{e}_i \cdot (\mathbf{e}_j \times \mathbf{e}_k). \tag{1.30}$$

Hence if i, j and k are all different, then $\mathbf{e}_j \times \mathbf{e}_k = \pm\mathbf{e}_i$, the plus sign being taken if i, j, k form a cyclic permutation of 1, 2, 3 (1, 2, 3 or 3, 1, 2 or 2, 3, 1) and the minus sign if they form an anticyclic permutation (3, 2, 1 or 2, 1, 3 or 1, 3, 2). Hence, from (1.30), $\epsilon_{ijk} = +1$ if i, j, k are all different and cyclic, and $\epsilon_{ijk} = -1$ if i, j, k are all different and anticyclic. If $j = k$, then the cross-product is zero in (1.30) and consequently so is ϵ_{ijk}. If either j or k equals i then $\mathbf{e}_j \times \mathbf{e}_k$ is at right-angles to $\mathbf{e}_i$ and ϵ_{ijk} will again be zero. We have finally

$$\epsilon_{ijk} = \begin{cases} +1 & \text{if } i, j, k \text{ are a cyclic permutation of } 1, 2, 3, \\ -1 & \text{if } i, j, k \text{ are an anticyclic permutation of } 1, 2, 3, \\ 0 & \text{if any two (or all) indices are equal.} \end{cases} \tag{1.31}$$

Hence cyclically permuting the indices i, j, k leaves ϵ_{ijk} unaffected, whereas interchanging any two indices changes its sign:

$$\epsilon_{ijk} = \epsilon_{kij} = \epsilon_{jki}, \tag{1.32}$$

$$\epsilon_{jik} = -\epsilon_{ijk}. \tag{1.33}$$

1.4 Vector identities

Consider the vector product $\mathbf{c} = \mathbf{a} \times \mathbf{b}$. Then from (1.2)

$$\mathbf{c} = (a_i\mathbf{e}_i) \times (b_j\mathbf{e}_j) = a_ib_j(\mathbf{e}_i \times \mathbf{e}_j). \tag{1.34}$$

Now, in general, $\mathbf{c} = (\mathbf{c} \cdot \mathbf{e}_k)\mathbf{e}_k$ using (1.3). Hence substituting for $\mathbf{c}$ from (1.34) gives

$$\mathbf{c} = \mathbf{a} \times \mathbf{b} = (a_ib_j(\mathbf{e}_i \times \mathbf{e}_j) \cdot \mathbf{e}_k)\mathbf{e}_k \tag{1.35}$$

$$= a_ib_j\epsilon_{kij}\mathbf{e}_k, \tag{1.36}$$

using (1.30). In (1.36), summation over the indices i, j and k is implied so that

$$\mathbf{a} \times \mathbf{b} = a_ib_j\epsilon_{1ij}\mathbf{e}_1 + a_ib_j\epsilon_{2ij}\mathbf{e}_2 + a_ib_j\epsilon_{3ij}\mathbf{e}_3 \tag{1.37}$$

and the rth component of $\mathbf{a} \times \mathbf{b}$ is therefore

$$(\mathbf{a} \times \mathbf{b})_r = \epsilon_{rij}a_ib_j, \quad (r = 1, 2, 3). \tag{1.38}$$

For the scalar triple product

$$(\mathbf{a} \times \mathbf{b}) \cdot \mathbf{c} = (a_ib_j\mathbf{e}_i \times \mathbf{e}_j) \cdot c_k\mathbf{e}_k \tag{1.39}$$

$$= a_ib_jc_k\mathbf{e}_k \cdot (\mathbf{e}_i \times \mathbf{e}_j) \tag{1.40}$$

$$= \epsilon_{kij}a_ib_jc_k. \tag{1.41}$$

Using (1.32), we have finally

$$(\mathbf{a} \times \mathbf{b}) \cdot \mathbf{c} = \epsilon_{ijk}a_ib_jc_k. \tag{1.42}$$

The scalar triple product does not depend on the order of the dot and cross operations since $\mathbf{a} \cdot (\mathbf{b} \times \mathbf{c}) = (\mathbf{b} \times \mathbf{c}) \cdot \mathbf{a} = \epsilon_{pqr}b_pc_qa_r = \epsilon_{rpq}a_rb_pc_q = (\mathbf{a} \times \mathbf{b}) \cdot \mathbf{c}$.

We can use a result from vector algebra to derive an important identity involving the Kronecker delta and the alternating symbol. We have the standard result

$$(\mathbf{a} \times \mathbf{b}) \cdot (\mathbf{c} \times \mathbf{d}) = (\mathbf{a} \cdot \mathbf{c})(\mathbf{b} \cdot \mathbf{d}) - (\mathbf{a} \cdot \mathbf{d})(\mathbf{b} \cdot \mathbf{c}). \tag{1.43}$$

Putting $\mathbf{a} = \mathbf{e}_i$, $\mathbf{b} = \mathbf{e}_j$, $\mathbf{c} = \mathbf{e}_k$ and $\mathbf{d} = \mathbf{e}_l$, then

$$(\mathbf{e}_i \times \mathbf{e}_j) \cdot (\mathbf{e}_k \times \mathbf{e}_l) = (\mathbf{e}_i \cdot \mathbf{e}_k)(\mathbf{e}_j \cdot \mathbf{e}_l) - (\mathbf{e}_i \cdot \mathbf{e}_l)(\mathbf{e}_j \cdot \mathbf{e}_k). \tag{1.44}$$

Now since for any vector $\mathbf{A}$, $\mathbf{A} = (\mathbf{A} \cdot \mathbf{e}_m)\mathbf{e}_m$ (as in (1.3)),

$$(\mathbf{e}_i \times \mathbf{e}_j) = [(\mathbf{e}_i \times \mathbf{e}_j) \cdot \mathbf{e}_m]\mathbf{e}_m = \epsilon_{mij}\mathbf{e}_m = \epsilon_{ijm}\mathbf{e}_m. \tag{1.45}$$

Similarly

$$(\mathbf{e}_k \times \mathbf{e}_l) = \epsilon_{klp}\mathbf{e}_p. \tag{1.46}$$

Hence the left-hand side of (1.44) becomes

$$(\mathbf{e}_i \times \mathbf{e}_j) \cdot (\mathbf{e}_k \times \mathbf{e}_l) = \epsilon_{ijm}\epsilon_{klp}\mathbf{e}_m \cdot \mathbf{e}_p \tag{1.47}$$

$$= \epsilon_{ijm}\epsilon_{klp}\delta_{mp} \tag{1.48}$$

$$= \epsilon_{ijm}\epsilon_{klm}, \tag{1.49}$$

using (1.23) and (1.29). Substituting this into (1.44) and expressing all the dot-products on the right-hand side using (1.23), we have

$$\epsilon_{ijm}\epsilon_{klm} = \delta_{ik}\delta_{jl} - \delta_{il}\delta_{jk}. \tag{1.50}$$

Example 6 A matrix ϕ has elements

$$\phi_{ik} = n_i n_k + \epsilon_{ilk} n_l, \tag{1.51}$$

where n_i are the components of a unit vector. Show that the elements of the matrix ϕ^2 are given by

$$(\phi^2)_{ij} = 2n_i n_j - \delta_{ij}. \tag{1.52}$$

Now

$$(\phi^2)_{ij} = \phi_{ik}\phi_{kj} = (n_i n_k + \epsilon_{ilk} n_l)(n_k n_j + \epsilon_{kmj} n_m) \tag{1.53}$$

$$= n_i n_j n_k n_k + (\epsilon_{ilk} n_l n_k) n_j + (\epsilon_{kmj} n_k n_m) n_i + \epsilon_{ilk}\epsilon_{kmj} n_l n_m. \tag{1.54}$$

Since ϵ_{pqr} is an antisymmetric symbol under interchange of any two indices, quantities such as $\epsilon_{ilk} n_l n_k$ are zero because for any pair of values l and k, two terms (with opposite signs) result from the summations over l and k (for example, $\epsilon_{i12} n_1 n_2$ cancels with $\epsilon_{i21} n_2 n_1 = -\epsilon_{i12} n_1 n_2$). Hence, since $n_k n_k = 1$,

$$(\phi^2)_{ij} = n_i n_j + \epsilon_{ilk}\epsilon_{kmj} n_l n_m \tag{1.55}$$

$$= n_i n_j + \epsilon_{ilk}\epsilon_{mjk} n_l n_m. \tag{1.56}$$

Using (1.50), (1.56) becomes

$$(\phi^2)_{ij} = n_i n_j + (\delta_{im}\delta_{lj} - \delta_{ij}\delta_{lm}) n_l n_m \tag{1.57}$$

$$= n_i n_j + n_j n_i - \delta_{ij} n_l n_l = 2n_i n_j - \delta_{ij}, \tag{1.58}$$

as required. ◢

1.5 Vector operators

Defining the operator ∇_i by

$$\nabla_i = \frac{\partial}{\partial x_i} \tag{1.59}$$

and the vector operator $\boldsymbol{\nabla}$ by

$$\boldsymbol{\nabla} = \mathbf{e}_i \nabla_i, \tag{1.60}$$

the gradient of a scalar function $\phi = \phi(x_1, x_2, x_3)$ is

$$\text{grad}\,\phi = \boldsymbol{\nabla}\phi = \mathbf{e}_i \nabla_i \phi = \mathbf{e}_1 \frac{\partial \phi}{\partial x_1} + \mathbf{e}_2 \frac{\partial \phi}{\partial x_2} + \mathbf{e}_3 \frac{\partial \phi}{\partial x_3}. \tag{1.61}$$

We note that the quantities $\partial x_i / \partial x_s$ in Example 5 above can be written

$$\partial x_i / \partial x_s = \nabla_s x_i = \delta_{is}. \tag{1.62}$$

The divergence of a vector function $\mathbf{a}(x_1, x_2, x_3)$ is

$$\text{div}\,\mathbf{a} = \boldsymbol{\nabla} \cdot \mathbf{a} = \nabla_i a_i = \frac{\partial a_1}{\partial x_1} + \frac{\partial a_2}{\partial x_2} + \frac{\partial a_3}{\partial x_3}. \tag{1.63}$$

The curl of a vector function $\mathbf{a}(x_1, x_2, x_3)$ can be expressed, using (1.38), as

$$(\boldsymbol{\nabla} \times \mathbf{a})_i = \epsilon_{ijk} \nabla_j a_k = \epsilon_{ijk} \frac{\partial a_k}{\partial x_j}, \tag{1.64}$$

giving

$$\text{curl}\,\mathbf{a} = \boldsymbol{\nabla} \times \mathbf{a} = \mathbf{e}_i \epsilon_{ijk} \nabla_j a_k. \tag{1.65}$$

The curl of a vector (in cartesian coordinates) can easily be written down in terms of determinants as follows: if in (1.41) $\mathbf{a} \cdot (\mathbf{b} \times \mathbf{c})$ is written out in full using the definition (1.31) of ϵ_{ijk}, we find

$$\mathbf{a} \cdot (\mathbf{b} \times \mathbf{c}) = \epsilon_{ijk} a_i b_j c_k \tag{1.66}$$

$$= a_1 b_2 c_3 - a_1 b_3 c_2 + a_2 b_3 c_1 - a_2 b_1 c_3 + a_3 b_1 c_2 - a_3 b_2 c_1, \tag{1.67}$$

which can be written as the determinant

$$\mathbf{a} \cdot (\mathbf{b} \times \mathbf{c}) = \begin{vmatrix} a_1 & a_2 & a_3 \\ b_1 & b_2 & b_3 \\ c_1 & c_2 & c_3 \end{vmatrix}. \tag{1.68}$$

Hence

$$\boldsymbol{\nabla} \times \mathbf{a} = \epsilon_{ijk} \mathbf{e}_i \nabla_j a_k \tag{1.69}$$

$$= \begin{vmatrix} \mathbf{e}_1 & \mathbf{e}_2 & \mathbf{e}_3 \\ \nabla_1 & \nabla_2 & \nabla_3 \\ a_1 & a_2 & a_3 \end{vmatrix}. \tag{1.70}$$

Various identities involving the vector operator can be derived using the above results (and the summation convention). We illustrate this with three examples.

Example 7

$$\mathrm{div}(\mathbf{b}\phi) = \nabla_i(b_i\phi) \tag{1.71}$$
$$= \phi\nabla_i b_i + (b_i\nabla_i\phi) \tag{1.72}$$
$$= \phi \,\mathrm{div}\,\mathbf{b} + (\mathbf{b}\,.\,\boldsymbol{\nabla})\phi. \tag{1.73}$$

Example 8

$$\mathrm{curl}(\mathbf{b}\phi) = \mathbf{e}_i\epsilon_{ijk}\nabla_j(b_k\phi) \tag{1.74}$$
$$= \phi\mathbf{e}_i(\epsilon_{ijk}\nabla_j b_k) + \mathbf{e}_i\epsilon_{ijk}(\nabla_j\phi)b_k \tag{1.75}$$
$$= \phi \,\mathrm{curl}\,\mathbf{b} + \mathbf{e}_i\epsilon_{ijk}(\boldsymbol{\nabla}\phi)_j b_k \tag{1.76}$$
$$= \phi \,\mathrm{curl}\,\mathbf{b} + (\boldsymbol{\nabla}\phi)\times\mathbf{b}, \tag{1.77}$$

where in the last term we have used (1.38) with $\mathbf{a} = \boldsymbol{\nabla}\phi$.

Example 9

$$\mathrm{curl\,curl}\,\mathbf{A} = \mathbf{e}_i\epsilon_{ijk}\nabla_j(\mathrm{curl}\,\mathbf{A})_k \tag{1.78}$$
$$= \mathbf{e}_i\epsilon_{ijk}\nabla_j(\epsilon_{klm}\nabla_l A_m) \tag{1.79}$$
$$= \mathbf{e}_i\epsilon_{ijk}\epsilon_{klm}\nabla_l\nabla_j A_m \tag{1.80}$$
$$= \mathbf{e}_i\epsilon_{ijk}\epsilon_{lmk}\nabla_l\nabla_j A_m \tag{1.81}$$
$$= \mathbf{e}_i(\delta_{il}\delta_{jm} - \delta_{im}\delta_{jl})\nabla_l\nabla_j A_m \tag{1.82}$$
$$= \mathbf{e}_i\nabla_i(\nabla_m A_m) - \nabla_j\nabla_j\mathbf{e}_i A_i \tag{1.83}$$
$$= \mathrm{grad}(\mathrm{div}\,\mathbf{A}) - \nabla^2\mathbf{A}, \tag{1.84}$$

where

$$\nabla^2 = \nabla_j\nabla_j = \frac{\partial^2}{\partial x_1^2} + \frac{\partial^2}{\partial x_2^2} + \frac{\partial^2}{\partial x_3^2}.$$

1.6 Orthogonal coordinate systems

So far we have considered only cartesian coordinates x_1, x_2 and x_3. We will require, in later chapters, expressions for the operator div grad ($=\nabla^2$) in coordinate systems based on cylindrical and spherical polar coordinates.

Consider two points with cartesian coordinates (x_1, x_2, x_3) and $(x_1+dx_1, x_2+dx_2, x_3+dx_3)$, where dx_i are small displacements. The infinitesimal distance ds between these two points is given by

$$ds^2 = dx_1^2 + dx_2^2 + dx_3^2 = dx_i\, dx_i \tag{1.85}$$

(using the summation convention). We now transform to a new coordinate system, say q_1, q_2 and q_3, the x_i being functions of the q_i. If ds^2 can be written in the form

$$ds^2 = h_1^2\, dq_1^2 + h_2^2\, dq_2^2 + h_3^2\, dq_3^2 = h_i^2\, dq_i^2, \tag{1.86}$$

then the new coordinates form an orthogonal coordinate system. For cartesian coordinates $h_1 = h_2 = h_3 = 1$ and $q_i = x_i$ $(i = 1, 2, 3)$. We now give, without proof, expressions for the gradient, divergence and curl in the new coordinate system in terms of the quantities h_i.

If Φ is a *scalar* and $\mathbf{A} = \mathbf{e}_i A_i$ is a vector then

$$\text{grad}\, \Phi = \boldsymbol{\nabla}\Phi = \frac{\mathbf{e}_1}{h_1}\frac{\partial \Phi}{\partial q_1} + \frac{\mathbf{e}_2}{h_2}\frac{\partial \Phi}{\partial q_2} + \frac{\mathbf{e}_3}{h_3}\frac{\partial \Phi}{\partial q_3}, \tag{1.87}$$

$$\text{div}\, \mathbf{A} = \frac{1}{h_1 h_2 h_3}\left[\frac{\partial}{\partial q_1}(h_2 h_3 A_1) + \frac{\partial}{\partial q_2}(h_1 h_3 A_2) + \frac{\partial}{\partial q_3}(h_1 h_2 A_3)\right], \tag{1.88}$$

$$\text{curl}\, \mathbf{A} = \frac{1}{h_1 h_2 h_3}\begin{vmatrix} h_1\mathbf{e}_1 & h_2\mathbf{e}_2 & h_3\mathbf{e}_3 \\ \dfrac{\partial}{\partial q_1} & \dfrac{\partial}{\partial q_2} & \dfrac{\partial}{\partial q_3} \\ h_1 A_1 & h_2 A_2 & h_3 A_3 \end{vmatrix}. \tag{1.89}$$

It is important to realise that in the above expressions the vectors $\mathbf{e}_i$ are unit vectors which are directed along the three new coordinate axes and point in the direction of increasing coordinate values. Further

$$\nabla^2\Phi = \frac{1}{h_1 h_2 h_3}\left[\frac{\partial}{\partial q_1}\left(\frac{h_2 h_3}{h_1}\frac{\partial \Phi}{\partial q_1}\right) + \frac{\partial}{\partial q_2}\left(\frac{h_1 h_3}{h_2}\frac{\partial \Phi}{\partial q_2}\right) + \frac{\partial}{\partial q_3}\left(\frac{h_1 h_2}{h_3}\frac{\partial \Phi}{\partial q_3}\right)\right]. \tag{1.90}$$

We now specialise these results to two particular coordinate systems which are of importance in later chapters.

1. Cylindrical polar coordinates

We specify z, the distance of the point from the x_1, x_2 plane, and the polar coordinates ρ and ϕ of the projection of the point in this plane. Thus $q_1 = \rho$, $q_2 = \phi$, $q_3 = z$ (see Figure 1.2). The vectors $\mathbf{e}_i$ point in the directions of increasing ρ, ϕ and z. The relationships between cartesian and cylindrical polar coordinates are

$$x_1 = \rho \cos \phi, \quad x_2 = \rho \sin \phi, \quad x_3 = z. \tag{1.91}$$

Hence

$$dx_1 = \frac{\partial x_1}{\partial \rho} d\rho + \frac{\partial x_1}{\partial \phi} d\phi + \frac{\partial x_1}{\partial z} dz \tag{1.92}$$

$$= \cos \phi \, d\rho - \rho \sin \phi \, d\phi. \tag{1.93}$$

Similarly

$$dx_2 = \sin \phi \, d\rho + \rho \cos \phi \, d\phi, \tag{1.94}$$

$$dx_3 = dz. \tag{1.95}$$

Hence

$$ds^2 = dx_i \, dx_i = (\cos \phi \, d\rho - \rho \sin \phi \, d\phi)^2 + (\sin \phi \, d\rho + \rho \cos \phi \, d\phi)^2 + dz^2 \tag{1.96}$$

$$= d\rho^2 + \rho^2 \, d\phi^2 + dz^2. \tag{1.97}$$

This is therefore an orthogonal coordinate system with

$$h_1 = 1, \quad h_2 = \rho, \quad h_3 = 1. \tag{1.98}$$

Hence, from (1.87)–(1.90),

$$\nabla\Phi = \mathbf{e}_1 \frac{\partial \Phi}{\partial \rho} + \frac{\mathbf{e}_2}{\rho} \frac{\partial \Phi}{\partial \phi} + \mathbf{e}_3 \frac{\partial \Phi}{\partial z}, \tag{1.99}$$

$$\operatorname{div} \mathbf{A} = \frac{1}{\rho} \left[\frac{\partial}{\partial \rho} (\rho A_1) + \frac{\partial A_2}{\partial \phi} + \rho \frac{\partial A_3}{\partial z} \right], \tag{1.100}$$

$$\operatorname{curl} \mathbf{A} = \frac{1}{\rho} \begin{vmatrix} \mathbf{e}_1 & \rho \mathbf{e}_2 & \mathbf{e}_3 \\ \dfrac{\partial}{\partial \rho} & \dfrac{\partial}{\partial \phi} & \dfrac{\partial}{\partial z} \\ A_1 & \rho A_2 & A_3 \end{vmatrix} \tag{1.101}$$

Figure 1.2

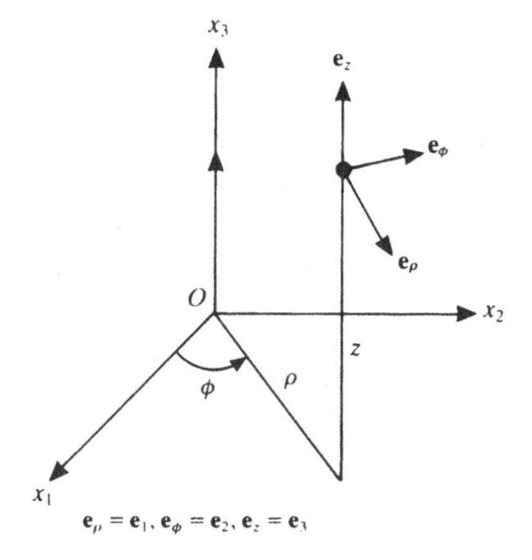

and

$$\nabla^2\Phi = \frac{1}{\rho}\left[\frac{\partial}{\partial \rho}\left(\rho\frac{\partial \Phi}{\partial \rho}\right) + \frac{1}{\rho}\frac{\partial^2\Phi}{\partial \phi^2} + \rho\frac{\partial^2\Phi}{\partial z^2}\right]. \tag{1.102}$$

2. Spherical polar coordinates

The coordinates specify the distance r of the point from the origin, and the 'latitude' θ and longitude ϕ, so that $q_1 = r$, $q_2 = \theta$, $q_3 = \phi$ (see Figure 1.3). The relationships between cartesian and spherical polar coordinates are

$$x_1 = r\sin\theta\cos\phi, \quad x_2 = r\sin\theta\sin\phi, \quad x_3 = r\cos\theta. \tag{1.103}$$

Hence

$$dx_1 = \frac{\partial x_1}{\partial r}dr + \frac{\partial x_1}{\partial \theta}d\theta + \frac{\partial x_1}{\partial \phi}d\phi \tag{1.104}$$

$$= \sin\theta\cos\phi\, dr + r\cos\theta\cos\phi\, d\theta - r\sin\theta\sin\phi\, d\phi. \tag{1.105}$$

Similarly

$$dx_2 = \sin\theta\sin\phi\, dr + r\cos\theta\sin\phi\, d\theta + r\sin\theta\cos\phi\, d\phi, \tag{1.106}$$

$$dx_3 = \cos\theta\, dr - r\sin\theta\, d\theta. \tag{1.107}$$

From (1.105)–(1.107)

$$ds^2 = dx_i\, dx_i = dr^2 + r^2\, d\theta^2 + r^2\sin^2\theta\, d\phi^2. \tag{1.108}$$

This is therefore an orthogonal coordinate system with

$$h_1 = 1, \quad h_2 = r, \quad h_3 = r\sin\theta. \tag{1.109}$$

Figure 1.3

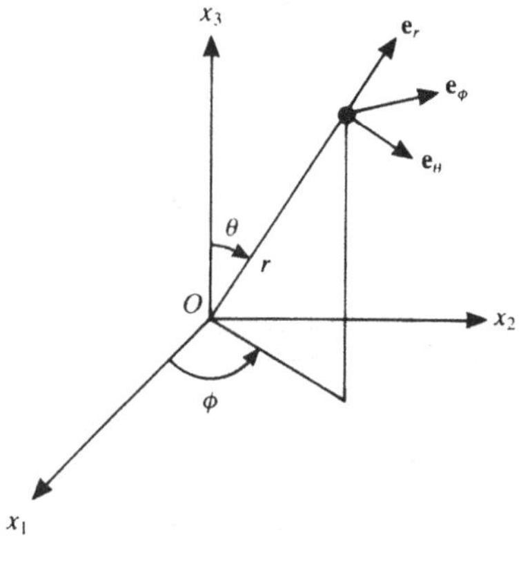

Hence, from (1.87)–(1.90),

$$\nabla\Phi = \mathbf{e}_1 \frac{\partial \Phi}{\partial r} + \frac{\mathbf{e}_2}{r} \frac{\partial \Phi}{\partial \theta} + \frac{\mathbf{e}_3}{r \sin\theta} \frac{\partial \Phi}{\partial \phi}, \tag{1.110}$$

$$\operatorname{div} \mathbf{A} = \frac{1}{r^2 \sin\theta} \left[\sin\theta \frac{\partial}{\partial r}(r^2 A_1) + r \frac{\partial}{\partial \theta}(\sin\theta\, A_2) + r \frac{\partial A_3}{\partial \phi} \right], \tag{1.111}$$

$$\operatorname{curl} \mathbf{A} = \frac{1}{r^2 \sin\theta} \begin{vmatrix} \mathbf{e}_1 & r\mathbf{e}_2 & r\sin\theta\, \mathbf{e}_3 \\ \dfrac{\partial}{\partial r} & \dfrac{\partial}{\partial \theta} & \dfrac{\partial}{\partial \phi} \\ A_1 & rA_2 & r\sin\theta\, A_3 \end{vmatrix} \tag{1.112}$$

and

$$\nabla^2\Phi = \frac{1}{r^2 \sin\theta} \left[\sin\theta \frac{\partial}{\partial r}\left(r^2 \frac{\partial \Phi}{\partial r}\right) + \frac{\partial}{\partial \theta}\left(\sin\theta \frac{\partial \Phi}{\partial \theta}\right) + \frac{1}{\sin\theta} \frac{\partial^2 \Phi}{\partial \phi^2} \right]. \tag{1.113}$$

1.7 General coordinate transformations

The tensor notation and the summation convention developed and used in the preceding sections is useful especially in the general theory of tensors. This is a large topic and here we only give a brief introduction.

Tensor theory is concerned with the way particular mathematical quantities transform under general coordinate transformations. Suppose we have an n-dimensional coordinate system (the x-system), a point in this space having coordinates $x_1, x_2, \ldots, x_n$. Now suppose that we carry out a transformation of these coordinates to a new coordinate system (the $\bar{x}$-system) where $\bar{x}_1$ is related to the x-coordinates by

$$\bar{x}_1 = f_1(x_1, x_2, \ldots, x_n). \tag{1.114}$$

Similarly,

$$\begin{aligned} \bar{x}_2 &= f_2(x_1, x_2, \ldots, x_n), \\ \vdots\ & \quad \vdots \\ \bar{x}_n &= f_n(x_1, x_2, \ldots, x_n), \end{aligned} \tag{1.115}$$

where $f_1, f_2, \ldots, f_n$ are functions specifying how the coordinates $\bar{x}_i$ are related to the coordinates x_i. Using a free index i, (1.114) and (1.115) may be written

$$\bar{x}_i = f_i(x_1, x_2, \ldots, x_n), \quad (i = 1, 2, \ldots, n), \tag{1.116}$$

or more compactly,

$$\bar{x}_i = f_i(x_s), \quad (s, i = 1, 2, \ldots, n). \tag{1.117}$$

For example, the transformations (1.91) from cylindrical polar coordinates (ρ, ϕ, z) to cartesian coordinates (x_1, x_2, x_3) are of this form, as are those in (1.103) from spherical polar coordinates (r, θ, ϕ) to cartesian coordinates (x_1, x_2, x_3).

Now assuming that the functions f_i are differentiable (up to whatever order we require), we can differentiate (1.117) to obtain

$$d\bar{x}_i = \frac{\partial f_i}{\partial x_1} dx_1 + \frac{\partial f_i}{\partial x_2} dx_2 + \ldots + \frac{\partial f_i}{\partial x_n} dx_n \tag{1.118}$$

$$= \frac{\partial f_i}{\partial x_j} dx_j, \tag{1.119}$$

or

$$d\bar{x}_i = \frac{\partial \bar{x}_i}{\partial x_j} dx_j, \tag{1.120}$$

where, using the summation convention, summation is over the index j (from 1 to n). Using (1.120), we have n equations for the $d\bar{x}_i$ in terms of the dx_i. In order that the dx_i may be uniquely expressed in terms of the $d\bar{x}_i$, we must require that the Jacobian determinant J (with element $\partial \bar{x}_i/\partial x_j$ in the ith row and jth column) satisfies

$$J = |\partial \bar{x}_i/\partial x_j| \neq 0. \tag{1.121}$$

When this condition is satisfied, the coordinate transformation (1.117) is called an allowable transformation. It then follows that we can solve (1.120) to give

$$dx_i = \frac{\partial x_i}{\partial \bar{x}_j} d\bar{x}_j, \quad (i = 1, 2, \ldots, n). \tag{1.122}$$

Geometrically, we can easily see the significance of these transformations. Suppose in the x-coordinate system P is a point with coordinates x_i and Q is a neighbouring point with coordinates $x_i + dx_i$. Then the quantities dx_i are the components of an infinitesimal vector joining P to Q. When the coordinate system is transformed to the $\bar{x}$-coordinates, P has coordinates $\bar{x}_i$ and Q has coordinates $\bar{x}_i + d\bar{x}_i$. Then (1.120) represents how the vector $d\bar{x}_i$ from P to Q in the $\bar{x}$-system is expressed in terms of the vector dx_i in the x-system.

1.8 Contravariant vectors

We now use the basic form of the transformation law (1.120) to define a contravariant vector. It is convention that the indices on the x_i and $\bar{x}_i$ in (1.120) are raised into superscript position so that we write (1.120) as

$$d\bar{x}^i = \frac{\partial \bar{x}^i}{\partial x^j} dx^j. \tag{1.123}$$

Definition A set of functions A^i of the coordinates is said to form a contravariant vector if under coordinate transformation from x^i to $\bar{x}^i$, A^i transforms to $\bar{A}^i$ where

$$\bar{A}^i = \frac{\partial \bar{x}^i}{\partial x^j} A^j, \quad (i = 1, 2, \ldots, n). \tag{1.124}$$

With this definition we see that the dx^i of (1.123) form a contravariant vector.

Quantities with indices in the superscript position are contravariant objects and a quantity with one free index (A^i) is called a tensor of rank one, the rank being equal to the number of free indices. Accordingly, if (1.124) holds, A^i is a contravariant tensor of rank one, a rank one object of this type being a vector.

1.9 Kronecker delta symbol

We now define the Kronecker delta symbol (see (1.23)) by putting one index into superscript position so that

$$\delta^i_k = \begin{cases} 1 & \text{if} \quad i = k, \\ 0 & \text{if} \quad i \neq k. \end{cases} \tag{1.125}$$

Then, using the summation convention,

$$\delta^i_j A^j = A^i, \tag{1.126}$$

$$\delta^i_j \delta^j_k = \delta^i_k, \tag{1.127}$$

$$\delta^i_i = n \quad \text{(the dimension of the space)}. \tag{1.128}$$

Using the Kronecker delta, we can now invert the transformation (1.124) for a contravariant vector as follows: multiplying (1.124) on both sides by $\partial x^k / \partial \bar{x}^i$ (and summing over i), we have

$$\frac{\partial x^k}{\partial \bar{x}^i} \bar{A}^i = \frac{\partial x^k}{\partial \bar{x}^i} \frac{\partial \bar{x}^i}{\partial x^j} A^j \tag{1.129}$$

$$= \frac{\partial x^k}{\partial x^j} A^j = \delta^k_j A^j = A^k. \tag{1.130}$$

Hence

$$A^k = \frac{\partial x^k}{\partial \bar{x}^i}\bar{A}^i \tag{1.131}$$

is the inverse transformation of (1.124) (provided that (1.121) holds).

1.10 Scalars and covariant vectors

Any quantity $\phi(x^i)$ (a function of the coordinates x^i) which has the same value under the transformation of coordinates from x^i to $\bar{x}^i$, in the sense that

$$\phi(x^i) = \bar{\phi}(\bar{x}^i), \tag{1.132}$$

is called a scalar quantity. We say that a scalar is invariant under coordinate transformations.

Now consider

$$A_i = \partial\phi/\partial x^i \tag{1.133}$$

and

$$\bar{A}_i = \partial\bar{\phi}/\partial\bar{x}^i. \tag{1.134}$$

Then by the chain rule of differentiation,

$$\frac{\partial\bar{\phi}}{\partial\bar{x}^i} = \frac{\partial\bar{\phi}}{\partial x^j}\frac{\partial x^j}{\partial\bar{x}^i} = \frac{\partial\phi}{\partial x^j}\frac{\partial x^j}{\partial\bar{x}^i}, \tag{1.135}$$

using (1.132) (summation over j is implied). Hence the objects A_i defined by (1.133) transform as

$$\bar{A}_i = \frac{\partial x^j}{\partial\bar{x}^i}A_j, \quad (i = 1, 2, \ldots, n). \tag{1.136}$$

Such an object is called a covariant vector. We note the difference between (1.124) and (1.136). However, if we transform from one rectangular cartesian coordinate system to another, linearly related to the first, it can be shown that there is no distinction between covariant and contravariant tensors. Such tensors are called cartesian tensors.

The inverse of (1.136) may be easily obtained, since

$$\frac{\partial\bar{x}^i}{\partial x^k}\bar{A}_i = \frac{\partial\bar{x}^i}{\partial x^k}\frac{\partial x^j}{\partial\bar{x}^i}A_j \tag{1.137}$$

$$= \delta^j_k A_j = A_k, \tag{1.138}$$

whence

$$A_k = \frac{\partial\bar{x}^i}{\partial x^k}\bar{A}_i. \tag{1.139}$$

Suppose A^i is a contravariant vector (or contravariant tensor of rank one) and B_i is a covariant vector (or covariant tensor of rank one). Then, using (1.124) and (1.136),

$$\bar{A}^i \bar{B}_i = \frac{\partial \bar{x}^i}{\partial x^s} A^s \frac{\partial x^p}{\partial \bar{x}^i} B_p, \tag{1.140}$$

where summation over both s and p is implied. Hence

$$\bar{A}^i \bar{B}_i = \delta^p_s A^s B_p = A^p B_p = A^i B_i, \tag{1.141}$$

since p is a dummy index and may be written as i. The quantity $\bar{A}^i \bar{B}_i$ is therefore equal to $A^i B_i$ under coordinate transformation and hence is a scalar called the scalar product of the vectors A^i and B_i. Sometimes this scalar is called the inner product of A^i and B_i, whereas the object $A^i B_j$ is not a scalar but a tensor of rank two and is called the outer product of A^i and B_j. In elementary discussions, a vector is defined as an object which has both magnitude and direction. Strictly speaking, a vector (or tensor) is defined in terms of its transformation properties as in (1.124) or (1.136).

Example 10 Show that the vector A^i with components

$$A^1 = x + y, \quad A^2 = y - x, \quad A^3 = 0 \tag{1.142}$$

has the same form under rotation about the z-axis.

Under rotation about the z-axis, the coordinates x, y and z transform to

$$\bar{x} = x \cos\theta + y \sin\theta, \tag{1.143}$$

$$\bar{y} = -x \sin\theta + y \cos\theta, \tag{1.144}$$

$$\bar{z} = z. \tag{1.145}$$

With

$$x^1 = x, \quad x^2 = y, \quad x^3 = z \tag{1.146}$$

and

$$\bar{x}^1 = \bar{x}, \quad \bar{x}^2 = \bar{y}, \quad \bar{x}^3 = \bar{z}, \tag{1.147}$$

we now use the transformation law (1.124) to obtain

$$\bar{A}^1 = \frac{\partial \bar{x}}{\partial x} A^1 + \frac{\partial \bar{x}}{\partial y} A^2 + \frac{\partial \bar{x}}{\partial z} A^3 \tag{1.148}$$

$$= (x + y)\cos\theta + (y - x)\sin\theta \tag{1.149}$$

$$= (x \cos\theta + y \sin\theta) + (-x \sin\theta + y \cos\theta) \tag{1.150}$$

$$= \bar{x} + \bar{y}. \tag{1.151}$$

Similarly

$$\bar{A}^2 = \bar{y} - \bar{x}, \quad \bar{A}^3 = 0. \tag{1.152}$$

Equations (1.151) and (1.152) are of the same form as (1.142), as required. ◢

1.11 Tensors of higher rank

Tensors of higher rank (meaning more than one free index) may be defined by a simple extension of the transformations (1.124) and (1.136), so that every superscript index transforms as (1.124) and every subscript as (1.136). For example, the objects with the following transformation laws are all tensors of rank two, but of different types:

$$\bar{T}^{ij} = \frac{\partial \bar{x}^i}{\partial x^k}\frac{\partial \bar{x}^j}{\partial x^l} T^{kl}, \quad \text{(contravariant, rank 2)} \tag{1.153}$$

$$\bar{T}^i{}_j = \frac{\partial \bar{x}^i}{\partial x^k}\frac{\partial x^l}{\partial \bar{x}^j} T^k{}_l, \quad \text{(mixed, rank 2)} \tag{1.154}$$

$$\bar{T}_{ij} = \frac{\partial x^k}{\partial \bar{x}^i}\frac{\partial x^l}{\partial \bar{x}^j} T_{kl}, \quad \text{(covariant, rank 2)} \tag{1.155}$$

where the mixed tensor has one contravariant and one covariant index. Tensors of higher rank may be defined in an obvious way.

The algebra of tensors is straightforward since we may readily define the sum and difference of two tensors using the basic transformation laws. For example,

$$\bar{A}^i \pm \bar{B}^i = \frac{\partial \bar{x}^i}{\partial x^j}(A^j \pm B^j), \tag{1.156}$$

so that if A^i and B^i are contravariant vectors their sum and difference are also contravariant vectors. Similarly, it can be seen from the appropriate transformation law that the sum and difference of two tensors of the same type are tensors of that type.

An important property of all tensor equations is that if the components of a tensor are zero in one coordinate system, then they remain zero in any other coordinate system (as can be seen from the transformation laws).

The calculus of tensors, which we omit here, involves another quantity called the affine connection which does not possess tensorial properties: it may have zero components in one coordinate system but non-zero components in another.

Finally, we consider the infinitesimal distance ds between two neighbouring points. In cartesian coordinates, ds^2 is given by (1.85). In a general coordinate system, we may write

$$ds^2 = g_{ij}\, dx^i\, dx^j, \tag{1.157}$$

where g_{ij} is a covariant rank two tensor. Expression (1.157) is known as the metric and the elements g_{ij} $(i, j = 1, 2, \ldots, n)$ are the metric coefficients. Consider a coordinate transformation from x^i to $\bar{x}^i$. Then using (1.123) and (1.155),

$$d\bar{s}^2 = \bar{g}_{ij}\, d\bar{x}^i\, d\bar{x}^j \tag{1.158}$$

$$= \frac{\partial x^k}{\partial \bar{x}^i} \frac{\partial x^l}{\partial \bar{x}^j} g_{kl} \frac{\partial \bar{x}^i}{\partial x^p} dx^p \frac{\partial \bar{x}^j}{\partial x^q} dx^q \tag{1.159}$$

$$= \frac{\partial x^k}{\partial \bar{x}^i} \frac{\partial \bar{x}^i}{\partial x^p} \frac{\partial x^l}{\partial \bar{x}^j} \frac{\partial \bar{x}^j}{\partial x^q} g_{kl}\, dx^p\, dx^q \tag{1.160}$$

$$= \delta^k_p \delta^l_q g_{kl}\, dx^p\, dx^q \tag{1.161}$$

$$= g_{kl}\, dx^k\, dx^l = ds^2. \tag{1.162}$$

Hence, the quantity ds is a scalar since it is invariant under a general coordinate transformation.

Tensor algebra and its associated calculus are important tools in the study of continuum mechanics and in the general theory of relativity.

Problems 1

1. Write out $a_{ik}x_i x_k$ in expanded form, assuming $a_{ik} = a_{ki}$, and $i, k = 1, 2, 3$.
2. Over which indices (if any) in the following expressions is summation implied?
   (i) $a_{ij}b_j$, (ii) $a_{ij}b_{jj}$, (iii) $a_{ij}b_{ji}$, (iv) $a_{ii} = b_{ii}$.
3. Find the values of $\delta_{ij}\delta_{ij}$, $\delta_{ij}\delta_{jk}\delta_{km}\delta_{im}$, $\epsilon_{jkl}A_k A_l$, and $\delta_{ik}\epsilon_{ikm}$, all indices ranging from 1 to 3.
4. Evaluate $\epsilon_{ikl}\epsilon_{jkl}$ and $\epsilon_{ijk}\epsilon_{ijk}$, all indices ranging from 1 to 3.
5. The object T_{ij} is related to S_{ij} by the relation

   $$T_{ij} = (\alpha\delta_{ij}\delta_{kl} + \beta\delta_{ik}\delta_{jl})S_{kl},$$

   where α and β are constants. Find S_{ij} in terms of T_{ij}, all indices ranging from 1 to 3. (Hint: first evaluate T_{ii}.)
6. The square matrices A and B have elements a_{ik} and b_{ik} respectively. Use the suffix notation to show that $(AB)^T = B^T A^T$,

where ^{T} denotes the transpose operation. If A is a 3×3 matrix show that $\epsilon_{ijk}a_{ip}a_{jq}a_{kr}=\epsilon_{pqr}\,|A|$, where $|A|$ denotes the determinant of A.

7. Show that the ith component of the vector $\mathbf{a}\times(\mathbf{b}\times\mathbf{a})$ may be written as $c_{ij}b_j$, where b_j are the components of $\mathbf{b}$. Determine the form of c_{ij} in terms of the components of $\mathbf{a}$ and show that it is a symmetric object.
8. Use the suffix notation to show that
   (i) $\operatorname{div}(\mathbf{A}\times\mathbf{B})=\mathbf{B}\,.\operatorname{curl}\mathbf{A}-\mathbf{A}\,.\operatorname{curl}\mathbf{B}$,
   (ii) $\operatorname{curl}(\mathbf{A}\times\mathbf{B})=\mathbf{A}\operatorname{div}\mathbf{B}-\mathbf{B}\operatorname{div}\mathbf{A}+(\mathbf{B}\,.\nabla)\mathbf{A}-(\mathbf{A}\,.\nabla)\mathbf{B}$,
   where $\mathbf{A}$ and $\mathbf{B}$ are vectors.
9. If $f_k=x_ix_j\epsilon_{ijk}+x_ix_ix_k$, show that

$$\partial f_k/\partial x_s=2x_sx_k+x_ix_i\delta_{ks}.$$

Find also $\partial^2 f_k/\partial x_r\,\partial x_s$, and deduce that $\partial^2 f_k/\partial x_s^2=2x_k+4x_s\delta_{ks}$, no summation over s being implied. Verify the last result directly for the cases $k=1$, $s=1$, and $k=1$, $s=2$.

10. If $A_k=x_nx_nx_mx_mx_k$, show that

$$\partial A_k/\partial x_p=\alpha x_px_kx_mx_m+x_nx_nx_mx_m\delta_{kp},$$

and determine the constant α. Show also that

$$\partial A_k/\partial x_k=\beta x_nx_nx_mx_m$$

and determine β for the cases (i) when all indices range from 1 to 3, and (ii) when all indices range from 1 to 7.

11. Using the form

$$\operatorname{div}(f\operatorname{grad}\phi)=\frac{\partial}{\partial x_j}\left(f\frac{\partial\phi}{\partial x_j}\right)$$

and assuming that both f and ϕ are functions of $u=x_nx_n$ only, show that

$$\operatorname{div}(f\operatorname{grad}\phi)=4u\frac{d}{du}\left(f\frac{d\phi}{du}\right)+2f\frac{d\phi}{du}\,\delta_{jj}.$$

12. Verify that, if $\phi(x_i)$ satisfies the equation $\nabla^2\phi+K^2\phi=0$, where K is a constant, then the second-order tensor

$$T_{ik}=(\nabla_i\nabla_k-\delta_{ik}\nabla^2)\phi,$$

where $\nabla_i\equiv\partial/\partial x_i$, is a solution of

$$\epsilon_{ijk}\epsilon_{ilm}\nabla_j\nabla_mT_{kp}-K^2T_{lp}=0.$$

13. The symmetric tensor g_{ik} is related to the three-index object Γ_{ijk} which is symmetric in its last two indices (that is, $\Gamma_{ijk} = \Gamma_{ikj}$) by the relation

$$\partial g_{ik} / \partial x_m - g_{is}\Gamma_{skm} - g_{sk}\Gamma_{sim} = 0.$$

Show that

$$g_{ms}\Gamma_{sik} = \frac{1}{2}\left(\frac{\partial g_{im}}{\partial x_k} + \frac{\partial g_{km}}{\partial x_i} - \frac{\partial g_{ik}}{\partial x_m}\right).$$

2
Special functions

2.1 Origins

This chapter is principally concerned with some of the special functions which occur in the solution of differential equations both ordinary and partial. These functions are more complicated than those met in elementary mathematical methods (for example, the sine, cosine, exponential and logarithmic functions) and often arise in the solutions of linear second-order differential equations of the form

$$p(x)\frac{d^2y}{dx^2}+q(x)\frac{dy}{dx}+r(x)y=0. \tag{2.1}$$

For example, the equation

$$x^2\frac{d^2y}{dx^2}+x\frac{dy}{dx}+(x^2-\nu^2)y=0, \tag{2.2}$$

where ν is a constant, is called Bessel's equation, the solutions of which are the Bessel functions. We discuss the solution of this equation in detail in Section 2.5 and derive various properties of the Bessel functions in order to illustrate the general methods of solution. These methods may be employed to derive the properties of other special functions and, rather than repeating the analysis, we will simply list some of their important properties for reference purposes and for use in later chapters. For an extensive account of special functions and their tabulated values, the reader should consult a standard reference work[†]. By building up an extensive list of such

[†] M. Abramowitz and I. A. Stegun, *Handbook of Mathematical Functions* (Dover, New York, 1964).

special functions and their properties, we may regard a problem or equation as being solved if its solution can be expressed in terms of these functions.

Besides arising as solutions of differential equations, other special functions are defined in terms of integrals. These occur in some of the analysis which follows in later chapters. The next three sections deal briefly with these functions.

2.2 The gamma-function

We define the gamma-function $\Gamma(x)$ by

$$\Gamma(x) = \int_0^\infty t^{x-1}e^{-t}\,dt, \tag{2.3}$$

where, in order for the integral to converge, $x > 0$. Now consider $\Gamma(x+1)$. Integrating by parts, we have

$$\Gamma(x+1) = \int_0^\infty t^x e^{-t}\,dt = [-t^x e^{-t}]_0^\infty + x\int_0^\infty t^{x-1}e^{-t}\,dt. \tag{2.4}$$

In (2.4), the integrated term is zero because at the upper limit the exponential dominates the t^x term, while at the lower limit t^x is zero since $x > 0$. Hence, using (2.3),

$$\Gamma(x+1) = x\int_0^\infty t^{x-1}e^{-t}\,dt = x\Gamma(x). \tag{2.5}$$

Equation (2.5) is a recurrence relation which connects $\Gamma(x+1)$ with $\Gamma(x)$. Now suppose that $x = n$, where $n \geqslant 1$ is an integer. Then

$$\Gamma(n+1) = n\Gamma(n) = n(n-1)\Gamma(n-1) = \ldots = n!\,\Gamma(1). \tag{2.6}$$

But

$$\Gamma(1) = \int_0^\infty e^{-t}\,dt = 1, \tag{2.7}$$

so that

$$\Gamma(n+1) = n! \tag{2.8}$$

Sometimes $\Gamma(x+1)$ is denoted by $x!$ and referred to as the factorial function. Hence we may define $0! = \Gamma(1) = 1$. Another useful value to calculate is

$$\Gamma(\tfrac{1}{2}) = \int_0^\infty t^{-\frac{1}{2}}e^{-t}\,dt = 2\int_0^\infty e^{-u^2}\,du = \sqrt{\pi}. \tag{2.9}$$

Here we have transformed the variable of integration to u, where $t = u^2$, and used the standard result

$$\int_0^\infty e^{-au^2}\,du = \tfrac{1}{2}\sqrt{\left(\frac{\pi}{a}\right)}, \quad (a > 0). \tag{2.10}$$

Hence, from equations (2.5) and (2.9),

$$\Gamma(\tfrac{3}{2}) = \tfrac{1}{2}\Gamma(\tfrac{1}{2}) = \tfrac{1}{2}\sqrt{\pi}, \tag{2.11}$$

$$\Gamma(\tfrac{5}{2}) = \tfrac{3}{2}\Gamma(\tfrac{3}{2}) = \tfrac{3}{4}\sqrt{\pi}, \tag{2.12}$$

and so on for other half-integer arguments.

The recurrence relation (2.5) may be used to define the gamma-function for negative values of its argument. Taking

$$\Gamma(x) = \frac{\Gamma(x+1)}{x} \tag{2.13}$$

we have

$$\Gamma(-\tfrac{1}{2}) = \frac{\Gamma(\tfrac{1}{2})}{-\tfrac{1}{2}} = -2\Gamma(\tfrac{1}{2}) = -2\sqrt{\pi}, \tag{2.14}$$

$$\Gamma(-\tfrac{3}{2}) = \frac{\Gamma(-\tfrac{1}{2})}{-\tfrac{3}{2}} = \tfrac{4}{3}\sqrt{\pi}. \tag{2.15}$$

Figure 2.1

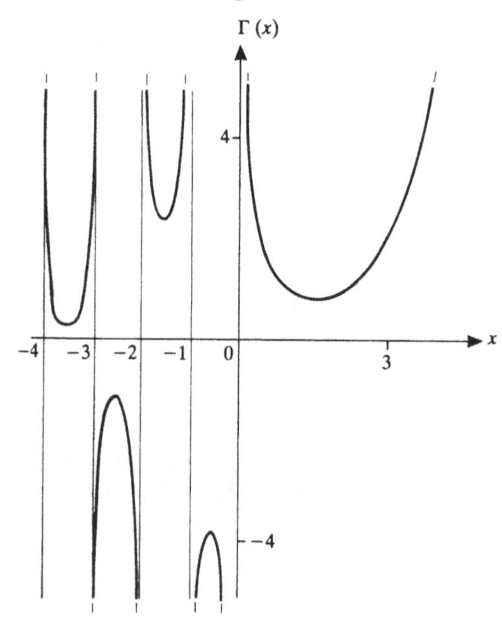

From equation (2.13), we see that $\Gamma(x)$ becomes infinite at $x = 0$ and hence also at $x = -1, -2, -3, \ldots$ as shown in Figure 2.1. Using the gamma-function, it is possible to evaluate various integrals, as shown by the following two examples.

Example 1 Consider

$$I_1 = \int_0^1 [\ln(1/x)]^{\frac{3}{2}}\, dx. \tag{2.16}$$

Putting $x = e^{-t}$, we find

$$I_1 = \int_0^\infty t^{\frac{3}{2}} e^{-t}\, dt = \Gamma(\tfrac{5}{2}) = \tfrac{3}{4}\sqrt{\pi}, \tag{2.17}$$

using (2.12). ◢

Example 2 Consider

$$I_2 = \int_0^\infty x^{\frac{1}{2}} e^{-x^3}\, dx. \tag{2.18}$$

Putting $x^3 = y$, we find

$$I_2 = \tfrac{1}{3}\int_0^\infty y^{-\frac{1}{2}} e^{-y}\, dy = \tfrac{1}{3}\Gamma(\tfrac{1}{2}) = \tfrac{1}{3}\sqrt{\pi}, \tag{2.19}$$

using (2.9). ◢

Other properties of the gamma-function are also of interest. Differentiating (2.3) with respect to x we have

$$\frac{d\Gamma(x)}{dx} = \Gamma'(x) = \int_0^\infty \frac{d}{dx}(t^x)\frac{e^{-t}}{t}\, dt. \tag{2.20}$$

Now

$$\frac{d}{dx}(t^x) = \frac{d}{dx}(e^{x\ln t}) = (\ln t)e^{x\ln t} = t^x \ln t. \tag{2.21}$$

Hence, from (2.20),

$$\Gamma'(x) = \int_0^\infty t^{x-1}(\ln t)e^{-t}\, dt. \tag{2.22}$$

Evaluating this at $x = 1$ gives

$$\Gamma'(1) = \int_0^\infty e^{-t}\ln t\, dt = -\gamma, \tag{2.23}$$

where γ is called Euler's constant and has the value (to four decimal places) 0·5772. This value can be obtained either by numerical integration in (2.22) or from the alternative definition

$$\gamma = \lim_{n\to\infty}\left(1+\tfrac{1}{2}+\tfrac{1}{3}+\tfrac{1}{4}+\ldots+\frac{1}{n}-\ln n\right). \tag{2.24}$$

A property of the gamma-function, which we give without proof, is

$$\Gamma(x)\Gamma(1-x)=\pi/\sin(\pi x) \tag{2.25}$$

which is known as the reflection formula.

Related to the gamma-function is the integral

$$\mathrm{B}(m, n)=\int_0^1 x^{m-1}(1-x)^{n-1}\,dx, \tag{2.26}$$

with $m>0$, $n>0$. This is called the beta-function. It is straightforward to show (see reference on page 22) that

$$\mathrm{B}(m, n)=\frac{\Gamma(m)\Gamma(n)}{\Gamma(m+n)}. \tag{2.27}$$

An alternative form of $\mathrm{B}(m, n)$ can be found from equation (2.26) by letting $x=\sin^2\theta$. Then, since $dx=2\sin\theta\cos\theta\,d\theta$, we have

$$\mathrm{B}(m, n)=2\int_0^{\pi/2}\sin^{2m-1}\theta\cos^{2n-1}\theta\,d\theta. \tag{2.28}$$

Example 3 Consider

$$I=\int_0^{\pi/2}\sqrt{(\tan\theta)}\,d\theta. \tag{2.29}$$

Then

$$I=\int_0^{\pi/2}\sin^{\frac{1}{2}}\theta\cos^{-\frac{1}{2}}\theta\,d\theta=\tfrac{1}{2}\mathrm{B}(\tfrac{3}{4},\tfrac{1}{4}). \tag{2.30}$$

Using (2.27), and $\Gamma(1)=1$, this can be written

$$I=\tfrac{1}{2}\Gamma(\tfrac{3}{4})\Gamma(\tfrac{1}{4})=\frac{\pi}{2\sin(\pi/4)}=\frac{\pi}{\sqrt{2}}, \tag{2.31}$$

where we have used (2.25) with $x=\frac{1}{4}$. ◢

The incomplete gamma-function $\Gamma(x, a)$ is defined by

$$\Gamma(x, a)=\int_a^\infty t^{x-1}\mathrm{e}^{-t}\,dt, \tag{2.32}$$

where $a \geqslant 0$, $x > 0$. This can be written as

$$\Gamma(x, a) = \int_0^\infty t^{x-1} e^{-t}\, dt - \int_0^a t^{x-1} e^{-t}\, dt \tag{2.33}$$

$$= \Gamma(x) - \int_0^a t^{x-1} e^{-t}\, dt. \tag{2.34}$$

Finally we obtain an approximation to $\Gamma(x+1)$ for large x known as Stirling's formula. Consider, from (2.3),

$$\Gamma(x+1) = \int_0^\infty t^x e^{-t}\, dt. \tag{2.35}$$

We change the variable of integration from t to τ, where $t = x + \tau\sqrt{x}$. Then

$$\Gamma(x+1) = \int_{-\sqrt{x}}^\infty (x + \tau\sqrt{x})^x e^{-(x+\tau\sqrt{x})} \sqrt{x}\, d\tau. \tag{2.36}$$

This can be written

$$\frac{\Gamma(x+1)}{e^{-x} x^{x+\frac{1}{2}}} = \int_{-\sqrt{x}}^\infty e^{-\tau\sqrt{x}} \left(1 + \frac{\tau}{\sqrt{x}}\right)^x d\tau \tag{2.37}$$

$$= \int_{-\sqrt{x}}^\infty \exp\left[-\tau\sqrt{x} + x \ln\left(1 + \frac{\tau}{\sqrt{x}}\right)\right] d\tau. \tag{2.38}$$

Writing (2.38) as the sum of two integrals and replacing the logarithm term by its MacLaurin expansion in $-\sqrt{x} < \tau < \sqrt{x}$, we have

$$\frac{\Gamma(x+1)}{e^{-x} x^{x+\frac{1}{2}}} = \int_{-\sqrt{x}}^{\sqrt{x}} \exp\left[-\tau\sqrt{x} + x\left(\frac{\tau}{\sqrt{x}} - \frac{\tau^2}{2x} + \ldots\right)\right] d\tau$$

$$+ \int_{\sqrt{x}}^\infty \exp\left[-\tau\sqrt{x} + x \ln\left(1 + \frac{\tau}{\sqrt{x}}\right)\right] d\tau. \tag{2.39}$$

For large x, the second integral has a vanishingly small range of integration, whilst the first integral gives

$$\frac{\Gamma(x+1)}{e^{-x} x^{x+\frac{1}{2}}} \approx \int_{-\infty}^\infty e^{-\tau^2/2}\, d\tau = \sqrt{(2\pi)}, \tag{2.40}$$

where we have again used the standard result (2.10). When $x = n$, a positive integer, $\Gamma(n+1) = n!$ so that, from (2.40), an approximation to $n!$ for large n is

$$n! \approx \sqrt{(2\pi)} e^{-n} n^{n+\frac{1}{2}}. \tag{2.41}$$

This result is surprisingly accurate even for modest values of n. For example, if $n = 3$, $n! = 6$ compared with an approximate value of 5·836 from (2.41). Similarly, when $n = 10$, $n! = 3\,628\,800$ compared with the approximate value from (2.41) of 3 598 695·618, which is within 1% of the exact value.

2.3 The exponential integral and related functions

The three integrals $E_1(x)$, $Ci(x)$, and $Si(x)$ are defined for $x > 0$ by

$$E_1(x) = \int_x^\infty \frac{e^{-t}}{t}\,dt, \tag{2.42}$$

$$Ci(x) = -\int_x^\infty \frac{\cos t}{t}\,dt, \tag{2.43}$$

$$Si(x) = \int_0^x \frac{\sin t}{t}\,dt. \tag{2.44}$$

The graphs of these functions are shown in Figures 2.2(a) and 2.2(b). These integrals arise in the Laplace inversion of logarithmic functions, in particular

$$\mathscr{L}\{E_1(at)\} = \frac{1}{s}\ln\left(\frac{s+a}{a}\right), \tag{2.45}$$

$$\mathscr{L}\{Ci(at)\} = -\frac{1}{2s}\ln\left(\frac{s^2+a^2}{a^2}\right), \tag{2.46}$$

where $\mathscr{L}$ denotes the Laplace transform (see Chapter 7). It is important to realise that the definitions (2.42)–(2.44) vary slightly between texts. $Ci(x)$ is sometimes defined without the minus sign. There also exists a function $Ei(x)$ which must not be confused with $E_1(x)$, and the function $si(x)$ defined by

$$si(x) = Si(x) - \frac{\pi}{2} = -\int_x^\infty \frac{\sin t}{t}\,dt. \tag{2.47}$$

2.4 The error function

The error function, denoted by erf x, is defined by the integral

$$\operatorname{erf} x = \frac{2}{\sqrt{\pi}}\int_0^x e^{-u^2}\,du. \tag{2.48}$$

Again using the standard result (2.10), we see that $\operatorname{erf} x \to 1$ as $x \to \infty$. The graph of the error function is shown in Figure 2.3.

The complement of the error function, denoted $\operatorname{erfc} x$, is defined by

$$\operatorname{erfc} x = 1 - \operatorname{erf} x = \frac{2}{\sqrt{\pi}} \int_x^\infty e^{-u^2}\, du \tag{2.49}$$

(see also Figure 2.3). Both of these functions arise from the solution of partial differential equations and are connected with the Laplace transforms of particular functions (see Chapters 7 and 8).

It is relatively easy for these functions to generate what is called an

Figure 2.2

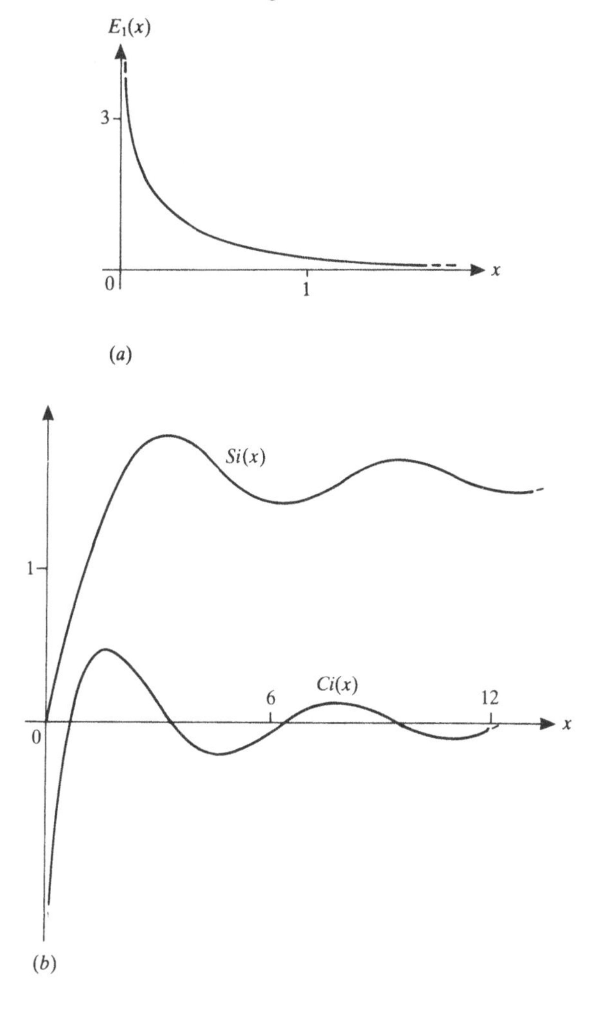

asymptotic series. We first give a definition of such a series. The series

$$S(x) = a_0 + \frac{a_1}{x} + \frac{a_2}{x^2} + \ldots + \frac{a_n}{x^n} + \ldots, \tag{2.50}$$

where $a_0, a_1, a_2, \ldots$ are constants, is said to be an asymptotic series of the function $f(x)$, written $f(x) \sim S(x)$, provided that, for any n, the error involved in truncating the series at the term a_n/x^n tends to zero faster than a_n/x^n as $x \to \infty$. In other words we require

$$\lim_{x\to\infty} x^n[f(x) - S(x)] = 0. \tag{2.51}$$

Asymptotic series may be added, multiplied and integrated to obtain asymptotic series for the sum, product and integral of the corresponding functions. It is instructive to point out the difference between a convergent series and an asymptotic series: a convergent series tends to the corresponding function $f(x)$ as $n \to \infty$ for a given x, whereas an asymptotic series tends to $f(x)$ as $x \to \infty$ for a given n (that is, truncating the series at the term a_n/x^n).

Consider now equation (2.49) for the complement of the error function and rewrite the integral

$$\int_x^\infty e^{-u^2}\,du = \int_x^\infty \left(-\frac{1}{2u}\right)(-2ue^{-u^2})\,du. \tag{2.52}$$

Integrating the right-hand side of (2.52) by parts we have

$$\int_x^\infty e^{-u^2}\,du = \left[\left(-\frac{1}{2u}\right)e^{-u^2}\right]_x^\infty - \int_x^\infty \frac{1}{2u^2}e^{-u^2}\,du \tag{2.53}$$

$$= \frac{e^{-x^2}}{2x} - \int_x^\infty \frac{1}{2u^2}e^{-u^2}\,du. \tag{2.54}$$

Figure 2.3

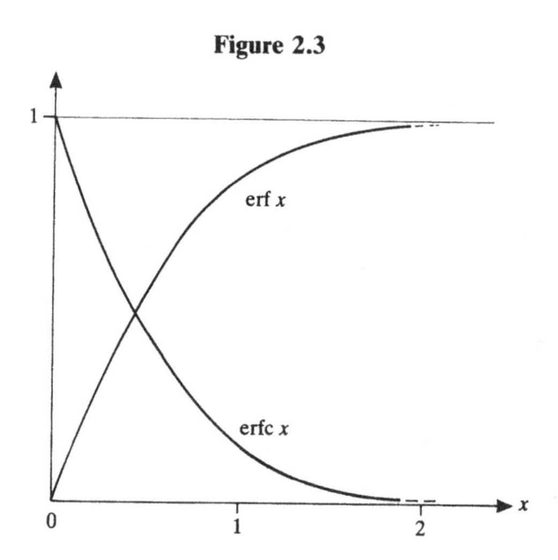

The last integral may be integrated by parts as follows:

$$\int_x^\infty \frac{1}{2u^2}\,\mathrm{e}^{-u^2}\,du = \int_x^\infty \left(-\frac{1}{4u^3}\right)(-2u\mathrm{e}^{-u^2})\,du \tag{2.55}$$

$$= \left[-\frac{1}{4u^3}\,\mathrm{e}^{-u^2}\right]_x^\infty - \tfrac{3}{4}\int_x^\infty \frac{1}{u^4}\,\mathrm{e}^{-u^2}\,du \tag{2.56}$$

$$= \frac{\mathrm{e}^{-x^2}}{4x^3} - \tfrac{3}{4}\int_x^\infty \frac{1}{u^4}\,\mathrm{e}^{-u^2}\,du. \tag{2.57}$$

Continuing in this way, we eventually find

$$\int_x^\infty \mathrm{e}^{-u^2}\,du \sim \frac{\mathrm{e}^{-x^2}}{2x}\left[1 - \frac{1}{2x^2} + \frac{1.3}{(2x^2)^2} - \frac{1.3.5}{(2x^2)^3} + \ldots + (-1)^n\frac{1.3.5.\ldots.(2n-1)}{(2x^2)^n} + R_{n+1}\right], \tag{2.58}$$

where

$$R_{n+1} = (-1)^{n+1}1.3.5.\ldots.(2n+1)2x\mathrm{e}^{x^2}\int_x^\infty \frac{1}{(2u^2)^{n+1}}\,\mathrm{e}^{-u^2}\,du. \tag{2.59}$$

Equation (2.58) is the asymptotic series for $\frac{1}{2}\sqrt{\pi}\,\mathrm{erfc}\,x$ since the ratio of consecutive terms is $(2n-1)/2x^2$ which tends to zero as $x \to \infty$ for a given n. However, for a given x, the series diverges as $n \to \infty$. By choosing a suitable n, the resulting finite series is a good approximation to the integral in (2.58). Accordingly, using (2.49) and (2.58), it is seen that the behaviour of $\mathrm{erfc}\,x$ for large x and to first order in $1/x$ is

$$\mathrm{erfc}\,x \sim \frac{\mathrm{e}^{-x^2}}{x\sqrt{\pi}}. \tag{2.60}$$

2.5 The Bessel function

The Bessel equation of order ν (where $\nu \geqslant 0$ is a constant) is (see (2.2))

$$x^2\frac{d^2y}{dx^2} + x\frac{dy}{dx} + (x^2 - \nu^2)y = 0. \tag{2.61}$$

We attempt a series solution of this equation by using the Frobenius series

$$y = x^m(a_0 + a_1x + a_2x^2 + \ldots) = \sum_{r=0}^\infty a_r x^{m+r}, \tag{2.62}$$

where, without loss of generality, we shall assume $a_0 \neq 0$. (If a_0 were zero, we could redefine m to preserve the form of the series (2.62).)

The conditions under which a series solution of this kind is valid are discussed in Chapter 4. In (2.62), m and the a_r are constants which must be found so that (2.62) satisfies (2.61). By this method, we will find two possible values of m, giving two solutions of (2.61). The solutions may not be independent if the values of m are identical or differ by an integer (see Chapter 4) but the method will always generate one solution. The method for finding a second solution in this case will be discussed later. The Frobenius series is more general than a Taylor or MacLaurin series since m may be found to have non-integer values and the resulting Frobenius series will then contain non-integer powers of x. This contrasts with the MacLaurin series which only contains integer powers of x.

Differentiating (2.62) we obtain

$$\frac{dy}{dx} = \sum_{r=0}^{\infty} a_r(m+r)x^{m+r-1} \tag{2.63}$$

and

$$\frac{d^2y}{dx^2} = \sum_{r=0}^{\infty} a_r(m+r)(m+r-1)x^{m+r-2}. \tag{2.64}$$

Inserting these into (2.61) we have

$$x^2 \sum_{r=0}^{\infty} a_r(m+r)(m+r-1)x^{m+r-2} + x \sum_{r=0}^{\infty} a_r(m+r)x^{m+r-1} + (x^2 - \nu^2) \sum_{r=0}^{\infty} a_r x^{m+r} = 0. \tag{2.65}$$

Collecting together like powers of x gives

$$\sum_{r=0}^{\infty} [a_r(m+r)(m+r-1) + a_r(m+r) - \nu^2 a_r]x^{m+r} + \sum_{r=0}^{\infty} a_r x^{m+r+2} = 0. \tag{2.66}$$

This simplifies to

$$\sum_{r=0}^{\infty} a_r[(m+r)^2 - \nu^2]x^{m+r} + \sum_{r=0}^{\infty} a_r x^{m+r+2} = 0. \tag{2.67}$$

Expanding the first summation by writing out the first two terms explicitly we have

$$a_0(m^2 - \nu^2)x^m + a_1[(m+1)^2 - \nu^2]x^{m+1} + \sum_{r=2}^{\infty} a_r[(m+r)^2 - \nu^2]x^{m+r} + \sum_{r=0}^{\infty} a_r x^{m+r+2} = 0. \tag{2.68}$$

The first summation contains powers of x from x^{2+m} upwards, so that putting $r = s + 2$ in this sum gives

$$\sum_{r=2}^{\infty} a_r[(m+r)^2 - v^2]x^{m+r} = \sum_{s=0}^{\infty} a_{s+2}[(m+s+2)^2 - v^2]x^{m+s+2} \tag{2.69}$$

$$= \sum_{r=0}^{\infty} a_{r+2}[(m+r+2)^2 - v^2]x^{m+r+2}, \tag{2.70}$$

where the dummy index has been changed from s to r. Hence (2.68) can be written with the two summations combined as

$$a_0(m^2 - v^2)x^m + a_1[(m+1)^2 - v^2]x^{m+1}$$
$$+ \sum_{r=0}^{\infty} \{a_{r+2}[(m+r+2)^2 - v^2] + a_r\}x^{m+r+2} = 0. \tag{2.71}$$

Now, since $a_0 \neq 0$, we must have $m^2 - v^2 = 0$ in order that the sole term in x^m vanishes. Hence

$$m = \pm v. \tag{2.72}$$

Likewise, the second term is the only one in x^{m+1} so that

$$a_1[(m+1)^2 - v^2] = a_1(1 \pm 2v) = 0. \tag{2.73}$$

Hence $a_1 = 0$ unless $v = \frac{1}{2}$, but as mentioned earlier, this case will have values of m which differ by an integer and so may need to be considered separately. Further, from (2.71), the coefficient of the general term in x^{m+r+2} must vanish giving

$$a_{r+2} = -\frac{a_r}{(m+r+2)^2 - v^2} \tag{2.74}$$

for $r = 0, 1, 2, \ldots$.

We now consider the two possible values of the constant m separately.

Case $m = v$ In this case (2.74) becomes

$$a_{r+2} = -\frac{a_r}{(v+r+2)^2 - v^2} = -\frac{a_r}{(r+2)(2v+r+2)}, \tag{2.75}$$

so that

$$a_2 = -\frac{a_0}{2(2v+2)}, \quad a_4 = -\frac{a_2}{4(2v+4)} = \frac{a_0}{2\,.\,4(2v+2)(2v+4)}, \tag{2.76}$$

and so on for all values of r. The final result for $y(x)$ is, from (2.62),

$$y(x) = a_0 x^\nu \left[1 - \frac{x^2}{2(2\nu+2)} + \frac{x^4}{2 . 4(2\nu+2)(2\nu+4)} - \frac{x^6}{2 . 4 . 6(2\nu+2)(2\nu+4)(2\nu+6)} + \dots \right]. \quad (2.77)$$

To define the Bessel function of order ν uniquely it is conventional to choose a specific value of a_0. By taking

$$a_0 = 1/2^\nu \Gamma(\nu+1) \quad (2.78)$$

and using the recurrence relation (2.5) for the gamma-function, $\Gamma(\nu+1) = \nu\Gamma(\nu)$, we may write the series (2.77) as

$$y(x) = J_\nu(x) = \sum_{r=0}^{\infty} \frac{(-1)^r}{r!\,\Gamma(\nu+r+1)} \left(\frac{x}{2}\right)^{\nu+2r}, \quad (2.79)$$

where $J_\nu(x)$ is the Bessel function of order ν and $0!$ is unity by (2.8). When $\nu = n$, where n is a positive integer or zero, then $\Gamma(n+r+1) = (n+r)!$ from (2.8), and (2.79) becomes

$$y(x) = J_n(x) = \sum_{r=0}^{\infty} \frac{(-1)^r}{r!\,(n+r)!} \left(\frac{x}{2}\right)^{n+2r}. \quad (2.80)$$

In particular

$$J_0(x) = 1 - \frac{x^2}{2^2} + \frac{x^4}{2^2 . 4^2} - \frac{x^6}{2^2 . 4^2 . 6^2} + \dots, \quad (2.81)$$

$$J_1(x) = \frac{x}{2} - \frac{x^3}{2^2 . 4} + \frac{x^5}{2^2 . 4^2 . 6} - \frac{x^7}{2^2 . 4^2 . 6^2 . 8} + \dots, \quad (2.82)$$

from which we see that

$$dJ_0(x)/dx = -J_1(x). \quad (2.83)$$

From the series (2.80) it is easy to see that all the $J_n(x)$ satisfy $J_n(0) = 0$ for n a positive integer, whereas if $n = 0$, $J_0(0) = 1$. The graphs of the first two, $J_0(x)$ and $J_1(x)$, are shown in Figure 2.4.

We now have to consider the second possible value of m obtained in (2.72).

Case $m = -\nu$ In this case we obtain the solution by simply changing the sign of ν in (2.77) and (2.79). If $\nu = 0, 1, 2, \dots$ then changing the sign of ν will result in diverging coefficients in (2.77). For example, (2.77) with ν replaced by $-\nu$ contains coefficients $1/(-2\nu+4)$

which diverge if $\nu = 2$. Further, the definition of a_0 in (2.78) would be $2^\nu/\Gamma(-\nu+1)$ which is zero if ν is an integer. Hence the cases $\nu = 0, 1, 2, \ldots$ must be treated separately. If then $\nu \neq 0, 1, 2, \ldots$ we have

$$y(x) = J_{-\nu}(x) = \sum_{r=0}^{\infty} \frac{(-1)^r}{r!\,\Gamma(-\nu+r+1)} \left(\frac{x}{2}\right)^{-\nu+2r}. \tag{2.84}$$

The two solutions (2.79) and (2.84) are independent since (2.79) contains only positive powers of x whereas (2.84) contains both positive and negative powers of x. The first is zero at $x = 0$ (except for the case $\nu = 0$) whilst the second is infinite at $x = 0$.

If $\nu = n$, an integer, then (2.73) still holds and $a_1 = 0$. From (2.71), the recurrence relation is

$$a_{r+2}(r+2)[2n-(r+2)] = a_r. \tag{2.85}$$

Writing out these in full gives

$$\begin{aligned} a_2 \,.\, 2(2n-2) &= a_0, \\ a_4 \,.\, 4(2n-4) &= a_2, \\ a_6 \,.\, 6(2n-6) &= a_4, \end{aligned} \tag{2.86}$$

and similar equations for $a_8, a_{10}, \ldots$. Only the even values of r need to be considered since (2.85) with $a_1 = 0$ implies that all a_r with r odd are zero. From (2.86) we see that if, say, $n = 3$ then $a_4 = 0$ which, working up the list, implies a_2 and hence a_0 are zero. In general then all the a_r preceding that for $r = 2n$ are zero. We then have from (2.85)

$$a_{2n+2} = -\frac{a_{2n}}{2(2n+2)}, \tag{2.87}$$

$$a_{2n+4} = -\frac{a_{2n+2}}{4(2n+4)} = \frac{a_{2n}}{2\,.\,4(2n+2)(2n+4)}, \tag{2.88}$$

Figure 2.4

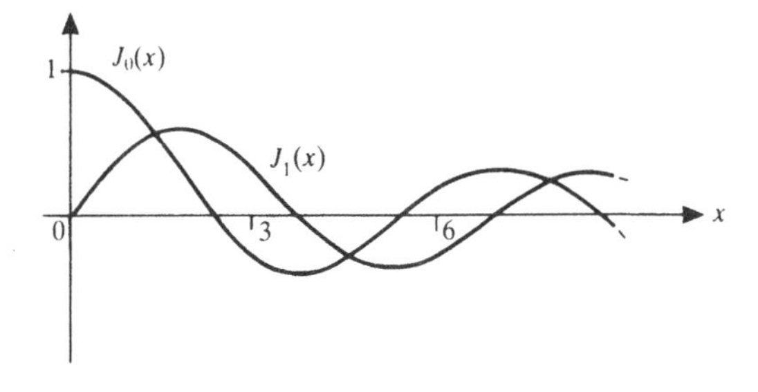

and so on. The series for $y(x)$ is

$$y(x) = a_{2n}x^{-n}\left[x^{2n} - \frac{x^{2n+2}}{2(2n+2)} + \frac{x^{2n+4}}{2.4(2n+2)(2n+4)} - \ldots\right] \tag{2.89}$$

$$= a_{2n}x^{n}\left[1 - \frac{x^{2}}{2(2n+2)} + \frac{x^{4}}{2.4(2n+2)(2n+4)} - \ldots\right] \tag{2.90}$$

and hence

$$J_{-n}(x) = a_{2n} \sum_{r=0}^{\infty} \frac{(-1)^r}{r!\,(n+r)!} \left(\frac{x}{2}\right)^{n+2r}. \tag{2.91}$$

From (2.80) we see that $J_n(x)$ and $J_{-n}(x)$ are linearly dependent. We normally choose a_{2n} so that

$$J_{-n}(x) = (-1)^n J_n(x). \tag{2.92}$$

The general solution of the Bessel equation may be written as

$$y(x) = AJ_\nu(x) + BJ_{-\nu}(x), \tag{2.93}$$

where A and B are constants, only when ν is not an integer. When $\nu = n$ we need to obtain a second solution ($J_n(x)$ being one solution) using the method in the next section.

2.6 The Bessel function Y_ν

To find the second solution of the Bessel equation when $\nu = n$, we write

$$y(x) = u(x)J_n(x) \tag{2.94}$$

and substitute this into the Bessel equation (2.61). This gives

$$u(x^2J_n'' + xJ_n' + (x^2 - \nu^2)J_n) + 2x^2u_n'J_n' + x^2u''J_n + xu'J_n = 0, \tag{2.95}$$

where a prime denotes differentiation with respect to x. Since J_n satisfies Bessel's equation (2.61) the first bracket in (2.95) is zero. Consequently

$$\frac{u''}{u'} + 2\frac{J_n'}{J_n} + \frac{1}{x} = 0. \tag{2.96}$$

Integrating with respect to x gives

$$\ln u' + \ln J_n^2 + \ln x = \ln A, \tag{2.97}$$

where for convenience we have chosen the arbitrary constant to be $\ln A$. Hence

$$u' = A/xJ_n^2 \tag{2.98}$$

and therefore the second solution, often denoted by $Y_n(x)$, is

$$Y_n(x) = u(x)J_n(x) = AJ_n(x)\int \frac{dx}{xJ_n^2(x)}. \tag{2.99}$$

The general solution of the Bessel equation is then

$$y(x) = AJ_n(x)\int \frac{dx}{xJ_n^2(x)} + BJ_n(x), \tag{2.100}$$

where A and B are arbitrary constants.

When $n = 0$, for example, we have

$$Y_0(x) = AJ_0(x)\int \frac{dx}{xJ_0^2(x)}. \tag{2.101}$$

Inserting the series (2.81) for $J_0(x)$ and taking only the first two terms, we find for small x

$$Y_0(x) \sim AJ_0(x)\int \frac{1}{x}\left(1 - \frac{x^2}{4}\right)^{-2} dx. \tag{2.102}$$

Performing the integral gives

$$Y_0(x) \sim AJ_0(x)[\ln x + P(x)], \tag{2.103}$$

where $P(x)$ is a polynomial in x. Hence the second solution is logarithmically divergent at $x = 0$. In fact, all the $Y_n(x)$ are singular at $x = 0$. The graphs of the first two are shown in Figure 2.5.

Alternative forms exist for the function $Y_n(x)$. Consider the Neumann or Weber function defined by

$$Y_\nu(x) = \frac{\cos(\nu\pi)J_\nu(x) - J_{-\nu}(x)}{\sin(\nu\pi)}. \tag{2.104}$$

Figure 2.5

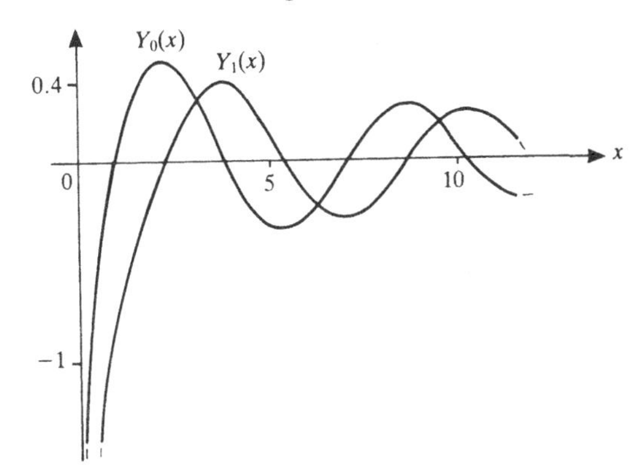

When $\nu \neq n$, this is just a linear combination of $J_\nu(x)$ and $J_{-\nu}(x)$ and so is a solution of Bessel's equation (independent of $J_\nu(x)$). When $\nu = n$, the numerator and denominator are both zero, so the limit as $\nu \to n$ must be taken using L'Hôpital's rule. Hence

$$Y_n(x) = \lim_{\nu \to n} \left\{ \frac{-\pi \sin(\nu\pi) J_\nu(x) + \left[\cos(\nu\pi) \dfrac{\partial J_\nu}{\partial \nu} - \dfrac{\partial J_{-\nu}}{\partial \nu}\right]}{\pi \cos(\nu\pi)} \right\} \tag{2.105}$$

$$= \frac{1}{\pi} \lim_{\nu \to n} \left[\frac{\partial J_\nu}{\partial \nu} - \sec(\nu\pi) \frac{\partial J_{-\nu}}{\partial \nu} \right] \tag{2.106}$$

$$= \frac{1}{\pi} \lim_{\nu \to n} \left[\frac{\partial J_\nu}{\partial \nu} - (-1)^n \frac{\partial J_{-\nu}}{\partial \nu} \right]. \tag{2.107}$$

To see why this is a solution of Bessel's equation consider the solutions J_ν and $J_{-\nu}$. These satisfy

$$x^2 \frac{d^2 J_\nu}{dx^2} + x \frac{dJ_\nu}{dx} + (x^2 - \nu^2) J_\nu = 0, \tag{2.108}$$

$$x^2 \frac{d^2 J_{-\nu}}{dx^2} + x \frac{dJ_{-\nu}}{dx} + (x^2 - \nu^2) J_{-\nu} = 0. \tag{2.109}$$

We differentiate each of these equations with respect to ν. Then multiplying the second by $(-1)^{n+1}$, where n is an integer, and adding it to the first, we obtain

$$x^2 \frac{d^2 u_\nu}{dx^2} + x \frac{du_\nu}{dx} + (x^2 - \nu^2) u_\nu = 2\nu[J_\nu - (-1)^n J_{-\nu}], \tag{2.110}$$

where

$$u_\nu = \frac{\partial J_\nu}{\partial \nu} - (-1)^n \frac{\partial J_{-\nu}}{\partial \nu}. \tag{2.111}$$

We now let $\nu \to n$. Then from (2.92), the right-hand side of (2.110) tends to zero and we find that

$$u_n = \lim_{\nu \to n} u_\nu \tag{2.112}$$

also satisfies Bessel's equation and is therefore the required second solution. Now from (2.107)

$$Y_n = \frac{1}{\pi} u_n \tag{2.113}$$

so that $Y_n(x)$ defined by (2.104) is also a solution of Bessel's equation.

2.7 Generating function

It is possible to generate Bessel functions of integral order as the coefficients of particular series. Consider the expansion of the functions $e^{xt/2}$ and $e^{-x/2t}$ in powers of t:

$$e^{xt/2} = 1 + \frac{xt}{2} + \frac{x^2t^2}{2!\,4} + \ldots = \sum_{r=0}^{\infty} \frac{1}{r!}\left(\frac{xt}{2}\right)^r, \tag{2.114}$$

$$e^{-x/2t} = 1 - \frac{x}{2t} + \frac{x^2}{2!\,4t^2} - \ldots = \sum_{s=0}^{\infty} \frac{(-1)^s}{s!}\left(\frac{x}{2t}\right)^s. \tag{2.115}$$

Taking the product of the two series gives

$$e^{\frac{1}{2}x(t-1/t)} = \sum_{r=0}^{\infty} \sum_{s=0}^{\infty} \frac{(-1)^s}{r!\,s!} (\tfrac{1}{2}x)^{r+s} t^{r-s}. \tag{2.116}$$

We now examine the coefficients of various powers of t. The coefficient of t^0 is obtained from (2.116) by putting $r = s$ and evaluating the sum of all such terms. The result is

$$1 - \frac{x^2}{2^2} + \frac{x^4}{2^2 \, . \, 4^2} - \ldots = J_0(x), \tag{2.117}$$

using (2.81). Likewise the coefficient of t^n is $J_n(x)$ and that of t^{-n} is $J_{-n}(x)$. Hence we may write

$$e^{\frac{1}{2}x(t-1/t)} = \sum_{n=-\infty}^{\infty} t^n J_n(x). \tag{2.118}$$

The function on the left-hand side of (2.118) is called the generating function for the Bessel functions and the series on the right is called a Laurent series (see Chapter 5) rather than a MacLaurin series, since the summation includes positive and negative powers of t. Various properties of the Bessel functions can be obtained from the generating function as follows.

1. Recurrence relations

(*a*) Differentiating (2.118) with respect to x gives

$$\tfrac{1}{2}(t - 1/t)e^{\frac{1}{2}x(t-1/t)} = \sum_{n=-\infty}^{\infty} t^n J_n'(x) \tag{2.119}$$

or, substituting for the generating function,

$$\tfrac{1}{2}(t - 1/t) \sum_{n=-\infty}^{\infty} t^n J_n(x) = \sum_{n=-\infty}^{\infty} t^n J_n'(x). \tag{2.120}$$

Hence

$$\tfrac{1}{2}\sum_{n=-\infty}^{\infty} t^{n+1}J_n(x) - \tfrac{1}{2}\sum_{n=-\infty}^{\infty} t^{n-1}J_n(x) = \sum_{n=-\infty}^{\infty} t^n J_n'(x). \qquad (2.121)$$

Comparing like powers of t on each side of (2.121) gives

$$\tfrac{1}{2}[J_{n-1}(x) - J_{n+1}(x)] = J_n'(x), \qquad (2.122)$$

whence

$$2J_n'(x) = J_{n-1}(x) - J_{n+1}(x). \qquad (2.123)$$

From this relation with $n = 0$, and using (2.92), we have

$$2J_0'(x) = J_{-1}(x) - J_1(x) = -2J_1(x) \qquad (2.124)$$

so that

$$J_0'(x) = -J_1(x) \qquad (2.125)$$

as in (2.83).

(*b*) Differentiating (2.118) with respect to t gives

$$\tfrac{1}{2}x(1 + 1/t^2)e^{\frac{1}{2}x(t-1/t)} = \sum_{n=-\infty}^{\infty} nt^{n-1}J_n(x) \qquad (2.126)$$

and hence

$$\tfrac{1}{2}x\sum_{n=-\infty}^{\infty} t^n J_n(x) + \tfrac{1}{2}x\sum_{n=-\infty}^{\infty} t^{n-2}J_n(x) = \sum_{n=-\infty}^{\infty} nt^{n-1}J_n(x). \qquad (2.127)$$

Comparing like powers of t on both sides gives

$$\tfrac{1}{2}xJ_{n-1}(x) + \tfrac{1}{2}xJ_{n+1}(x) = nJ_n(x), \qquad (2.128)$$

whence

$$\frac{2n}{x}J_n(x) = J_{n-1}(x) + J_{n+1}(x). \qquad (2.129)$$

(*c*) We have

$$\frac{d}{dx}[x^n J_n(x)] = x^n J_n'(x) + nx^{n-1}J_n(x) \qquad (2.130)$$

$$= \tfrac{1}{2}x^n[J_{n-1}(x) - J_{n+1}(x)] + nx^{n-1}J_n(x), \qquad (2.131)$$

using (2.123). Substituting for $J_{n+1}(x)$ from (2.129) gives

$$\frac{d}{dx}[x^n J_n(x)] = \tfrac{1}{2}x^n J_{n-1}(x) - \tfrac{1}{2}x^n\left[\frac{2n}{x}J_n(x) - J_{n-1}(x)\right] + nx^{n-1}J_n(x). \qquad (2.132)$$

Hence

$$\frac{d}{dx}[x^n J_n(x)] = x^n J_{n-1}(x). \qquad (2.133)$$

Example 4

$$I = \int x^2 J_1(x)\, dx = x^2 J_2(x) + C, \qquad (2.134)$$

where C is a constant, using (2.133). ◢

2. *Integral formula for Bessel functions*

In equation (2.118) we put $t = e^{i\theta}$ to simplify the following analysis. Since $t - 1/t = 2i \sin \theta$, we have from this equation

$$e^{ix \sin \theta} = \sum_{n=-\infty}^{\infty} t^n J_n(x) = \sum_{n=-\infty}^{\infty} e^{in\theta} J_n(x). \qquad (2.135)$$

Expanding the final summation and using the relation (2.92) we find

$$\begin{aligned} e^{ix \sin \theta} = J_0(x) &+ 2[J_2(x)\cos(2\theta) + J_4(x)\cos(4\theta) + \ldots] \\ &+ 2i[J_1(x)\sin\theta + J_3(x)\sin(3\theta) + \ldots]. \end{aligned} \qquad (2.136)$$

Taking the real and imaginary parts of (2.136) we have

$$\cos(x \sin \theta) = J_0(x) + 2\sum_{n=1}^{\infty} J_{2n}(x)\cos(2n\theta), \qquad (2.137)$$

$$\sin(x \sin \theta) = 2\sum_{n=0}^{\infty} J_{2n+1}(x)\sin[(2n+1)\theta]. \qquad (2.138)$$

Multiplying (2.137) by $\cos(m\theta)$ and (2.138) by $\sin(m\theta)$, where m is an integer, integrating with respect to θ from 0 to π and using the standard integrals

$$\int_0^{\pi} \cos(m\theta)\cos(n\theta)\, d\theta = \int_0^{\pi} \sin(m\theta)\sin(n\theta)\, d\theta = \tfrac{1}{2}\pi\delta_{mn}, \qquad (2.139)$$

(where δ_{mn} is the Kronecker delta symbol defined in (1.23)), we have

$$\frac{1}{\pi}\int_0^{\pi} \cos(x \sin\theta)\cos(m\theta)\, d\theta = \begin{cases} J_m(x) & \text{when } m \text{ is even} \\ 0 & \text{when } m \text{ is odd,} \end{cases} \qquad (2.140)$$

$$\frac{1}{\pi}\int_0^{\pi} \sin(x \sin\theta)\sin(m\theta)\, d\theta = \begin{cases} 0 & \text{when } m \text{ is even} \\ J_m(x) & \text{when } m \text{ is odd.} \end{cases} \qquad (2.141)$$

Hence, for all n, we have

$$J_n(x) = \frac{1}{\pi}\int_0^\pi [\cos(x \sin\theta)\cos(n\theta) + \sin(x\sin\theta)\sin(n\theta)]\, d\theta \tag{2.142}$$

$$= \frac{1}{\pi}\int_0^\pi \cos(n\theta - x\sin\theta)\, d\theta. \tag{2.143}$$

This integral form is useful in evaluating $J_n(x)$ numerically and can also be used to derive other relations. For example, we can obtain (2.92) from (2.143) by writing

$$J_{-n}(x) = \frac{1}{\pi}\int_0^\pi \cos(-n\theta - x\sin\theta)\, d\theta \tag{2.144}$$

$$= \frac{1}{\pi}\int_0^\pi \cos(n\theta + x\sin\theta)\, d\theta. \tag{2.145}$$

Now put $\theta = \pi - \phi$ in (2.145). Then

$$J_{-n}(x) = \frac{1}{\pi}\int_\pi^0 \cos[n(\pi - \phi) + x\sin\phi](-d\phi) \tag{2.146}$$

$$= \frac{1}{\pi}\int_0^\pi \cos(x\sin\phi - n\phi + n\pi)\, d\phi. \tag{2.147}$$

Now $\cos(A + n\pi) = (-1)^n \cos A$, so that

$$J_{-n}(x) = \frac{1}{\pi}(-1)^n \int_0^\pi \cos(n\phi - x\sin\phi)\, d\phi = (-1)^n J_n(x). \tag{2.148}$$

2.8 Modified Bessel functions

The general solution of the equation

$$x^2\frac{d^2y}{dx^2} + x\frac{dy}{dx} + (k^2x^2 - \nu^2)y = 0 \tag{2.149}$$

can be found from the standard form (2.61) of the Bessel equation by writing $t = kx$. Then (2.149) becomes

$$t^2\frac{d^2y}{dt^2} + t\frac{dy}{dt} + (t^2 - \nu^2)y = 0 \tag{2.150}$$

which has the general solution

$$y = AJ_\nu(t) + BY_\nu(t) = AJ_\nu(kx) + BY_\nu(kx). \tag{2.151}$$

An important case arises in particular physical problems when $k^2 = -1$. We then have the modified Bessel equation

$$x^2\frac{d^2y}{dx^2}+x\frac{dy}{dx}-(x^2+\nu^2)y=0 \tag{2.152}$$

with the general solution (since $k = \pm\mathrm{i}$)

$$y = AJ_\nu(\mathrm{i}x)+BY_\nu(\mathrm{i}x). \tag{2.153}$$

We now define a new function

$$I_\nu(x)=\mathrm{i}^{-\nu}J_\nu(\mathrm{i}x) \tag{2.154}$$

and use the series (2.79) for J_ν. Then

$$I_\nu(x)=\mathrm{i}^{-\nu}\sum_{r=0}^{\infty}\frac{(-1)^r}{r!\,\Gamma(\nu+r+1)}\left(\frac{\mathrm{i}x}{2}\right)^{\nu+2r} \tag{2.155}$$

$$=\sum_{r=0}^{\infty}\frac{1}{r!\,\Gamma(\nu+r+1)}\left(\frac{x}{2}\right)^{\nu+2r}, \tag{2.156}$$

which is a real function of x. Similar considerations apply to $K_\nu(x)$ which is the second solution of the modified Bessel equation (2.152). $I_\nu(x)$ and $K_\nu(x)$ are called modified Bessel functions and their properties can be obtained in a similar way to those of $J_\nu(x)$ and $Y_\nu(x)$. The main properties of these functions are given in the next section.

We note finally that the differential equation

$$d^2y/dx^2=(1-1/4x^2)y \tag{2.157}$$

can be solved in terms of the modified Bessel functions. Putting $y = x^{\frac{1}{2}}u$, we obtain the equation

$$x^2\frac{d^2u}{dx^2}+x\frac{du}{dx}-x^2u=0 \tag{2.158}$$

which, on comparison with (2.152), has solutions $I_0(x)$ and $K_0(x)$. Hence the general solution of (2.157) is

$$y(x)=Ax^{\frac{1}{2}}I_0(x)+Bx^{\frac{1}{2}}K_0(x). \tag{2.159}$$

Another equation which we shall meet in later chapters is

$$\frac{d^2y}{dx^2}=\left(\tfrac{1}{4}+\frac{\nu^2-\frac{1}{4}}{x^2}\right)y. \tag{2.160}$$

Again putting $y = x^{\frac{1}{2}}u$, we find

$$x^2\frac{d^2u}{dx^2}+x\frac{du}{dx}-(\tfrac{1}{4}x^2+\nu^2)u=0, \tag{2.161}$$

which the substitution $t = x/2$ converts into the modified Bessel equation in t:

$$t^2 \frac{d^2u}{dt^2} + t\frac{du}{dt} - (t^2 + \nu^2)u = 0. \tag{2.162}$$

The solutions, from (2.152), are therefore $I_\nu(t)$ and $K_\nu(t)$ so that the solution of (2.160) is

$$y(x) = Ax^{\frac{1}{2}}I_\nu(x/2) + Bx^{\frac{1}{2}}K_\nu(x/2). \tag{2.163}$$

2.9 Summary of the main properties of special functions

1. Bessel functions

Bessel's equation of order ν is

$$x^2\frac{d^2y}{dx^2} + x\frac{dy}{dx} + (x^2 - \nu^2)y = 0. \tag{2.164}$$

The independent solutions when ν is not zero or an integer n are $J_\nu(x)$ and $J_{-\nu}(x)$, where

$$J_\nu(x) = \sum_{r=0}^{\infty} \frac{(-1)^r}{r!\,\Gamma(\nu + r + 1)}\left(\frac{x}{2}\right)^{\nu+2r}. \tag{2.165}$$

When $\nu = n$,

$$J_{-n}(x) = (-1)^n J_n(x), \tag{2.166}$$

$$J_0(x) = 1 - \frac{x^2}{2^2} + \frac{x^4}{2^2 . 4^2} - \frac{x^6}{2^2 . 4^2 . 6^2} + \dots . \tag{2.167}$$

A second solution $Y_n(x)$ exists, in particular (see (2.103))

$$Y_0(x) = \frac{2}{\pi}\left[\ln\left(\frac{x}{2}\right) + \gamma\right]J_0(x) + \frac{x^2}{2^2} - \frac{x^4}{2^2 . 4^2}(1 + \tfrac{1}{2}) + \frac{x^6}{2^2 . 4^2 . 6^2}(1 + \tfrac{1}{2} + \tfrac{1}{3}) - \dots, \tag{2.168}$$

where γ is Euler's constant (see (2.23) and (2.24)).
The generating function is

$$e^{\frac{1}{2}x(t-1/t)} = \sum_{n=-\infty}^{\infty} t^n J_n(x). \tag{2.169}$$

Important recurrence relations are

$$2J_n'(x) = J_{n-1}(x) - J_{n+1}(x), \tag{2.170}$$

$$\frac{2n}{x} J_n(x) = J_{n-1}(x) + J_{n+1}(x), \tag{2.171}$$

(note $dJ_0(x)/dx = -J_1(x)$).
The integral form is

$$J_n(x) = \frac{1}{\pi} \int_0^\pi \cos(n\theta - x \sin\theta)\, d\theta. \tag{2.172}$$

The following are known as the Lommel integrals:

$$\int_0^l xJ_n(px)J_n(qx)\, dx = \frac{l}{q^2 - p^2} [pJ_n(ql)J_n'(pl) - qJ_n(pl)J_n'(ql)] \tag{2.173}$$

for $p \neq q$, and

$$\int_0^l xJ_n^2(px)\, dx = \frac{l^2}{2} \left[J_n'^2(pl) + \left(1 - \frac{n^2}{p^2 l^2}\right) J_n^2(pl) \right]. \tag{2.174}$$

2. *Modified Bessel functions*

The solutions of the equation

$$x^2 \frac{d^2 y}{dx^2} + x \frac{dy}{dx} - (x^2 + \nu^2) y = 0 \tag{2.175}$$

are the modified Bessel functions $I_\nu(x)$ and $K_\nu(x)$, where

$$I_\nu(x) = \mathrm{i}^{-\nu} J_\nu(\mathrm{i}x), \quad K_\nu(x) = \frac{\pi}{2} \mathrm{i}^{\nu+1} [\mathrm{i}Y_\nu(\mathrm{i}x) + J_\nu(\mathrm{i}x)]. \tag{2.176}$$

In particular

$$I_0(x) = J_0(\mathrm{i}x) = 1 + \frac{x^2}{2^2} + \frac{x^4}{2^2 . 4^2} + \frac{x^6}{2^2 . 4^2 . 6^2} + \ldots, \tag{2.177}$$

$$K_0(x) = -I_0(x)\left[\ln\left(\frac{x}{2}\right) + \gamma\right] + \frac{x^2}{4} + \ldots. \tag{2.178}$$

The graphs of $I_0(x)$ and $K_0(x)$ are given in Figure 2.6.

3. *Legendre polynomials*

Legendre's equation is

$$(1 - x^2) \frac{d^2 y}{dx^2} - 2x \frac{dy}{dx} + l(l+1) y = 0. \tag{2.179}$$

If l is not an integer, both solutions diverge at $x = \pm 1$. If $l = n = 0, 1, 2, \ldots$ the solutions are polynomials $P_n(x)$ called the Legendre polynomials, where

$$P_n(x) = \frac{1}{2^n n!} \frac{d^n}{dx^n} (x^2 - 1)^n. \tag{2.180}$$

In particular

$$P_0(x) = 1, \quad P_1(x) = x, \quad P_2(x) = \tfrac{1}{2}(3x^2 - 1). \tag{2.181}$$

The generating function is

$$\frac{1}{\sqrt{(1 - 2xt + t^2)}} = \sum_{n=0}^{\infty} t^n P_n(x). \tag{2.182}$$

Recurrence relations are

$$P_{n+1}(x) = \frac{2n+1}{n+1} x P_n(x) - \frac{n}{n+1} P_{n-1}(x), \tag{2.183}$$

$$P'_{n+1}(x) - P'_{n-1}(x) = (2n+1) P_n(x), \tag{2.184}$$

$$(1 - x^2) P'_n(x) = -nx P_n(x) + n P_{n-1}(x). \tag{2.185}$$

Figure 2.6

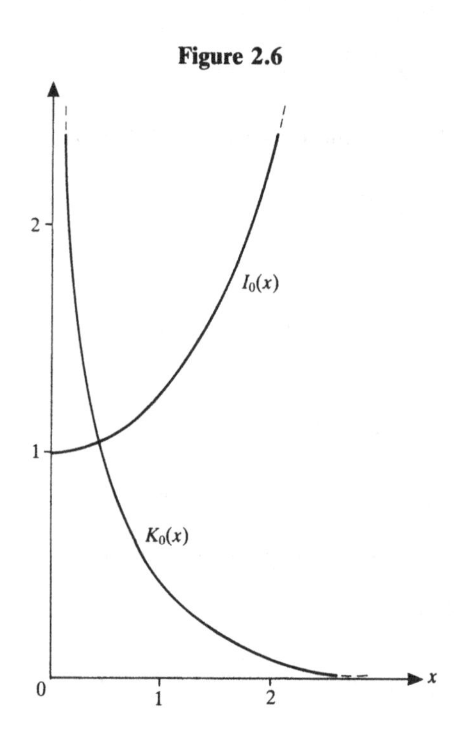

The integral form is

$$P_n(x) = \frac{1}{\pi}\int_0^{\pi} [x + \sqrt{(x^2+1)}\cos\theta]^n \, d\theta. \tag{2.186}$$

We also have the following integral:

$$\int_{-1}^{1} P_n(x)P_m(x)\, dx = \frac{2}{2n+1}\,\delta_{nm}. \tag{2.187}$$

Graphs of the first three Legendre polynomials are shown in Figure 2.7.
Further, the associated Legendre functions are defined by

$$P_n^m(x) = \frac{1}{2^n n!}(x^2-1)^{m/2}\frac{d^{m+n}}{dx^{m+n}}(x^2-1)^n, \tag{2.188}$$

where $0 \leqslant m \leqslant n$. In Chapter 8 we shall meet the equation

$$\frac{1}{\sin\theta}\frac{d}{d\theta}\left(\sin\theta\frac{dy}{d\theta}\right) = \left(\lambda + \frac{m^2}{\sin^2\theta}\right)y. \tag{2.189}$$

The solutions are bounded only if $\lambda = -n(n+1)$ and have the form

$$y = P_n^{|m|}(\cos\theta). \tag{2.190}$$

Figure 2.7

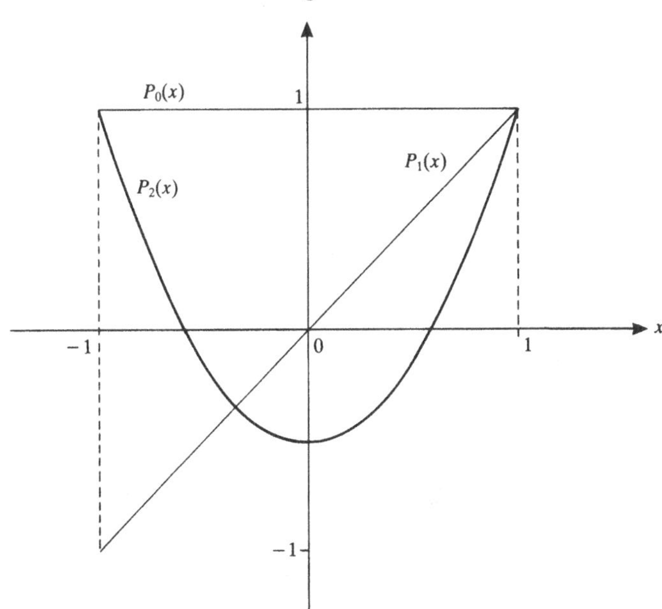

4. *Laguerre polynomials*

The Laguerre equation

$$x^2\frac{d^2y}{dx^2}+(1-x)\frac{dy}{dx}+ny=0, \tag{2.191}$$

where $n=0, 1, 2, \ldots$, has polynomial solutions $L_n(x)$, called the Laguerre polynomials, given by

$$L_n(x)=\frac{e^x}{n!}\frac{d^n}{dx^n}(x^n e^{-x}). \tag{2.192}$$

In particular

$$L_0(x)=1, \quad L_1(x)=1-x, \quad L_2(x)=\tfrac{1}{2}(x^2-4x+2). \tag{2.193}$$

The generating function is

$$\frac{e^{xt/(1-t)}}{1-t}=\sum_{n=0}^{\infty} t^n L_n(x). \tag{2.194}$$

A useful recurrence relation is

$$xL_n'(x)=nL_n(x)-nL_{n-1}(x). \tag{2.195}$$

We have the following integral:

$$\int_0^{\infty} e^{-x}L_m(x)L_n(x)\,dx=\delta_{mn}. \tag{2.196}$$

The associated Laguerre polynomials are defined by

$$L_n^m(x)=\frac{d^m}{dx^m}L_n(x) \tag{2.197}$$

for $n \geqslant m$ and satisfy

$$x\frac{d^2y}{dx^2}+(m+1-x)\frac{dy}{dx}+(n-m)y=0. \tag{2.198}$$

5. *Hermite polynomials*

The Hermite equation

$$\frac{d^2y}{dx^2}-2x\frac{dy}{dx}+2ny=0 \tag{2.199}$$

has polynomial solutions $H_n(x)$, called the Hermite polynomials, when $n=0, 1, 2, \ldots$, given by

$$H_n(x)=(-1)^n e^{x^2}\frac{d^n}{dx^n}(e^{-x^2}). \tag{2.200}$$

In particular

$$H_0(x)=1, \quad H_1(x)=2x, \quad H_2(x)=4x^2-2. \tag{2.201}$$

The generating function is

$$e^{2tx-t^2}=\sum_{n=0}^{\infty}\frac{1}{n!}t^nH_n(x). \tag{2.202}$$

Important recurrence relations are

$$H_{n+1}(x)-2xH_n(x)+2nH_{n-1}(x)=0, \tag{2.203}$$

$$H_n'(x)=2nH_{n-1}(x). \tag{2.204}$$

We also have the following integral:

$$\int_{-\infty}^{\infty} e^{-x^2}H_n(x)H_m(x)\,dx=2^n\sqrt{\pi}\,n!\,\delta_{nm}. \tag{2.205}$$

The Weber–Hermite function

$$y(x)=e^{-x^2/2}H_n(x) \tag{2.206}$$

satisfies the differential equation

$$\frac{d^2y}{dx^2}+(\lambda-x^2)y=0, \tag{2.207}$$

where

$$\lambda=2n+1. \tag{2.208}$$

If $\lambda \neq 2n+1$ in (2.207), then y is not finite as $x \to \pm\infty$.

Figure 2.8

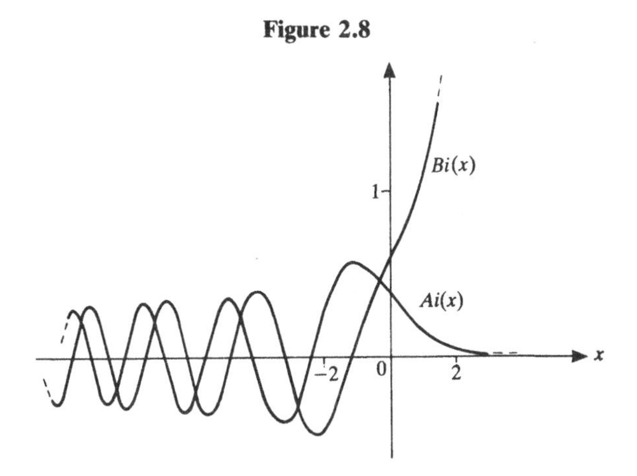

6. Airy functions

The Airy equation

$$\frac{d^2y}{dx^2}=xy \tag{2.209}$$

has solutions called the Airy functions, $Ai(x)$ and $Bi(x)$, where

$$Ai(x)=\sqrt{(x/3)}[I_{-\frac{1}{3}}(\tfrac{2}{3}x^{\frac{3}{2}})-I_{\frac{1}{3}}(\tfrac{2}{3}x^{\frac{3}{2}})], \tag{2.210}$$

$$Bi(x)=\sqrt{(x/3)}[I_{-\frac{1}{3}}(\tfrac{2}{3}x^{\frac{3}{2}})+I_{\frac{1}{3}}(\tfrac{2}{3}x^{\frac{3}{2}})]. \tag{2.211}$$

The integral form for $Ai(x)$ is

$$Ai(x)=\frac{1}{\pi}\int_0^\infty \cos(\tfrac{1}{3}t^3+xt)\,dt. \tag{2.212}$$

The graphs of $Ai(x)$ and $Bi(x)$ are given in Figure 2.8.

Problems 2

1. Use the gamma-function to evaluate

$$\text{(i)}\int_0^\infty x^4 e^{-x}\,dx,\quad \text{(ii)}\int_0^\infty \frac{e^{-at}}{\sqrt{t}}\,dt,\quad (a>0).$$

2. Show that

$$\int_0^1 \frac{dx}{\sqrt{(-\ln x)}}=\Gamma(\tfrac{1}{2}).$$

3. Show that

$$\int_0^\infty x^m e^{-kx^n}\,dx=\frac{1}{nk^{(m+1)/n}}\Gamma\left(\frac{m+1}{n}\right),\quad (k,m,n>0).$$

4. Evaluate $\int_0^1 dx/(1-x^3)^{\frac{1}{3}}$ using the beta-function.
5. Given

$$E_1(x)=\int_x^\infty \frac{e^{-t}}{t}\,dt,$$

show by integration by parts that

$$E_1(x)\sim\frac{e^{-x}}{x}\left(1-\frac{1!}{x}+\frac{2!}{x^2}-\frac{3!}{x^3}+\dots\right).$$

6. Use the beta-function to evaluate

$$\text{(i)}\int_0^{\pi/2}\sqrt{(\sin\theta)}\,d\theta,\quad \text{(ii)}\int_0^{\pi/2}\sqrt{(\cot\theta)}\,d\theta.$$

7. Show using equation (2.79) that

$$J_{\frac{1}{2}}(x)=\sqrt{\left(\frac{2}{\pi x}\right)}\sin x,\quad J_{-\frac{1}{2}}(x)=\sqrt{\left(\frac{2}{\pi x}\right)}\cos x$$

and that

$$J_{\frac{3}{2}}(x)=\sqrt{\left(\frac{2}{\pi x}\right)}\left(\frac{\sin x}{x}-\cos x\right),$$

$$J_{-\frac{3}{2}}(x)=\sqrt{\left(\frac{2}{\pi x}\right)}\left(-\sin x-\frac{\cos x}{x}\right).$$

8. Show, using (2.80) with $n=0$, that, for $\alpha>0$,

$$\int_0^\infty J_0(x)\mathrm{e}^{-\alpha x}\,dx=\frac{1}{\sqrt{(\alpha^2+1)}}.$$

Hence show that, in the limit $\alpha\to 0$,

$$\int_0^\infty J_0(x)\,dx=1.$$

9. Transform the equation

$$\frac{d^2y}{dx^2}+(\mathrm{e}^{2x}-\nu^2)y=0,$$

where ν is a constant, by writing $z=\mathrm{e}^x$. Hence obtain the general solution of this equation.

10. Show that Bessel's equation

$$x^2\frac{d^2y}{dx^2}+x\frac{dy}{dx}+(x^2-\nu^2)y=0$$

can be transformed into

$$\frac{d^2u}{dx^2}+\left(1-\frac{\nu^2-\frac{1}{4}}{x^2}\right)u=0,$$

where $y=x^{-\frac{1}{2}}u$. Obtain the general solution when $\nu=\pm\frac{1}{2}$. Use this result to show that for $x^2\gg\nu^2-\frac{1}{4}$

$$J_\nu(x)\sim\frac{1}{\sqrt{x}}(A\sin x+B\cos x).$$

11. Show that the general solution of

$$x\frac{d^2y}{dx^2}+y=0$$

is

$$y = \sqrt{x}[AJ_1(2\sqrt{x}) + BY_1(2\sqrt{x})].$$

12. Evaluate $\int J_1(x^{\frac{1}{3}})\,dx$, using Example 4.
13. Evaluate the following integrals using (2.180) and integration by parts:

$$\text{(i)} \int_0^1 xP_3(x)\,dx, \quad \text{(ii)} \int_{-1}^1 P_2(x)\ln(1-x)\,dx.$$

14. Show, using the standard integrals

$$\int_0^\infty \frac{\cos u}{u^p}\,du = \frac{\pi}{2\Gamma(p)\cos(\frac{1}{2}p\pi)}, \quad 0<p<1,$$

$$\int_0^\infty \frac{\sin u}{u^p}\,du = \frac{\pi}{2\Gamma(p)\sin(\frac{1}{2}p\pi)}, \quad 0<p<2,$$

and (2.212), that

$$\text{(i)}\ Ai(0) = \frac{1}{3^{\frac{2}{3}}\Gamma(\frac{2}{3})}, \quad \text{(ii)}\ Ai'(0) = -\frac{1}{3^{\frac{1}{3}}\Gamma(\frac{1}{3})}.$$

3
Non-linear ordinary differential equations

3.1 Introduction

The differential equations met in Chapter 2 were linear in the sense that they were special cases of the general nth order linear equation

$$\frac{d^n y}{dx^n} + f_1(x)\frac{d^{n-1}y}{dx^{n-1}} + \ldots + f_n(x)y = g(x), \tag{3.1}$$

where $f_1, f_2, \ldots, f_n, g$ are given functions of x. For such equations the Principle of Superposition applies: an arbitrary linear combination of individual solutions is also a solution.

Equations which cannot be written in the form of (3.1) are called non-linear and, for such equations, the Principle of Superposition does not apply. A typical first-order non-linear equation is

$$\left(\frac{dy}{dx}\right)^2 - y^2 = x^2, \tag{3.2}$$

whereas

$$\frac{d^2y}{dx^2} + y\frac{dy}{dx} + 6y = \sin x \tag{3.3}$$

is a second-order non-linear equation.

There is no general method of solving non-linear equations analytically and numerical procedures are frequently the only techniques available. However, some limited types can be solved analytically by special methods. These are discussed in subsequent sections.

3.2 Equations with separable variables

In general, the first-order equation

$$dy/dx = f(x, y)/g(x, y), \tag{3.4}$$

where $f(x, y)$ and $g(x, y)$ are given functions, will be non-linear. However, if $f(x, y)$ and $g(x, y)$ are separable so that $f(x, y) = X(x)Y(y)$ and $g(x, y) = U(x)V(y)$, then

$$dy/dx = F(x)/G(y), \tag{3.5}$$

where $F(x) = X(x)/U(x)$ and $G(y) = V(y)/Y(y)$. If we multiply (3.5) by $G(y)$ and integrate both sides with respect to x, we obtain

$$\int G(y)\,dy = \int F(x)\,dx + C \tag{3.6}$$

where C is a constant.

Example 1

$$\frac{dy}{dx} = \frac{\cos^2 y}{x}. \tag{3.7}$$

Separating the variables gives

$$\int \sec^2 y\,dy = \int \frac{dx}{x} + C, \tag{3.8}$$

where C is a constant. Carrying out the integrations and writing $C = \ln A$, where A is another constant, we obtain

$$\tan y = \ln x + \ln A \tag{3.9}$$

so that

$$y = \tan^{-1}[\ln(Ax)]. \blacktriangleleft \tag{3.10}$$

A case of a first-order non-linear equation which can be reduced to separable form is

$$dy/dx = K(y/x), \tag{3.11}$$

where K is a given function of y/x only. Writing $y = xu(x)$ gives

$$x\frac{du}{dx} + u = K(u), \tag{3.12}$$

or

$$\frac{du}{dx} = \frac{K(u) - u}{x}. \tag{3.13}$$

This equation is now of separable form.

Example 2

$$\frac{dy}{dx}=\frac{y}{x-y}. \tag{3.14}$$

Writing (3.14) in the form of (3.11), we have

$$\frac{dy}{dx}=\frac{y/x}{1-y/x}. \tag{3.15}$$

Putting $y=xu$ gives

$$x\frac{du}{dx}+u=\frac{u}{1-u}, \tag{3.16}$$

or

$$\frac{du}{dx}=\frac{u^2}{x(1-u)}. \tag{3.17}$$

Separating the variables leads to

$$\int\frac{1-u}{u^2}\,du=\int\frac{dx}{x}+\ln A, \tag{3.18}$$

where A is a constant. Integrating both sides, we have

$$-\frac{1}{u}-\ln u=\ln(Ax), \tag{3.19}$$

so that the relationship between x and y is, on substituting for $u=y/x$,

$$\frac{x}{y}+\ln(y/x)+\ln(Ax)=\frac{x}{y}+\ln(Ay)=0. \tag{3.20}$$

We note that in Example 2 we cannot determine y explicitly in terms of x. Although the separation of variables and integration can always, in principle, be carried out for equations of the type (3.5), there is no guarantee that y may be found explicitly in terms of x.

It may be possible, after a first integration, to reduce a second-order non-linear equation to a first-order separable type. We now give an example to illustrate this.

Example 3

$$x\frac{d^2y}{dx^2}=(y-1)\frac{dy}{dx}, \tag{3.21}$$

given that $y=1$ and $dy/dx=1$ at $x=0$.

Since

$$\frac{d}{dx}\left(x\frac{dy}{dx}\right) = x\frac{d^2y}{dx^2} + \frac{dy}{dx}, \tag{3.22}$$

then (3.21) can be written

$$\frac{d}{dx}\left(x\frac{dy}{dx}\right) - \frac{dy}{dx} = (y-1)\frac{dy}{dx}, \tag{3.23}$$

or, simplifying,

$$\frac{d}{dx}\left(x\frac{dy}{dx}\right) = y\frac{dy}{dx} = \frac{d}{dx}\left(\tfrac{1}{2}y^2\right). \tag{3.24}$$

Integrating with respect to x gives

$$x\frac{dy}{dx} = \tfrac{1}{2}y^2 + A, \tag{3.25}$$

where A is a constant. Applying the boundary conditions, we find that $A = -\frac{1}{2}$, so that (3.25) becomes

$$x\frac{dy}{dx} = \tfrac{1}{2}(y^2 - 1). \tag{3.26}$$

Separating the variables gives

$$2\int \frac{dy}{y^2 - 1} = \int \frac{dx}{x} + \ln B, \tag{3.27}$$

where B is a constant. The left-hand side can be integrated by expressing $(y^2 - 1)^{-1}$ in partial fractions. We have

$$\int \left(\frac{1}{y-1} - \frac{1}{y+1}\right) dy = \ln(Bx), \tag{3.28}$$

or

$$\ln\left(\frac{y-1}{y+1}\right) = \ln(Bx). \tag{3.29}$$

Hence

$$y - 1 = Bx(y+1), \tag{3.30}$$

or

$$y = \frac{1 + Bx}{1 - Bx}. \tag{3.31}$$

To find B, we again apply the boundary conditions. The condition $y = 1$ at $x = 0$ is satisfied by (3.31), and to satisfy the condition $dy/dx = 1$ at $x = 0$ we need to find dy/dx. From (3.31),

$$\frac{dy}{dx} = \frac{2B}{(1 - Bx)^2}. \tag{3.32}$$

At $x = 0$, $1 = 2B$, giving $B = \frac{1}{2}$. The complete solution, from (3.31), is therefore

$$y = \frac{2 + x}{2 - x}. \tag{3.33}$$

3.3 Equations reducible to linear form

In some cases a non-linear equation may be reduced to linear form and hence solved by standard techniques. The substitution $p = dy/dx$ may help to carry out this reduction.

Example 4 Consider

$$\left(y + \frac{dy}{dx}\right) \ln\left(y + \frac{dy}{dx}\right) + y = 0. \tag{3.34}$$

Differentiating with respect to x, we have

$$\left(\frac{dy}{dx} + \frac{d^2y}{dx^2}\right) \ln\left(y + \frac{dy}{dx}\right) + \frac{dy}{dx} + \frac{d^2y}{dx^2} + \frac{dy}{dx} = 0. \tag{3.35}$$

Eliminating the logarithm between (3.34) and (3.35) gives, on multiplying through by $y + dy/dx$,

$$\frac{dy}{dx}\frac{d^2y}{dx^2} + 2\left(\frac{dy}{dx}\right)^2 + y\frac{dy}{dx} = 0. \tag{3.36}$$

Hence either $dy/dx = 0$ or $d^2y/dx^2 + 2\,dy/dx + y = 0$, both of which are linear equations. The possible solutions are therefore

$$y = C \tag{3.37}$$

and

$$y = e^{-x}(Ax + B), \tag{3.38}$$

where A, B and C are constants.

We now determine the values of these constants so that (3.37) and (3.38) satisfy (3.34). Inserting $y = C$ into (3.34) gives

$$C \ln C + C = 0, \tag{3.39}$$

so that $C = 0$ or $C = e^{-1}$. In the case $C = 0$ the expression $C \ln C$ is undefined. Hence the solution is $C = e^{-1}$ giving

$$y = e^{-1}. \tag{3.40}$$

For $y = e^{-x}(Ax + B)$, $y + dy/dx = Ae^{-x}$. Inserting these into (3.34), we have

$$Ae^{-x} \ln(Ae^{-x}) + e^{-x}(Ax + B) = 0. \tag{3.41}$$

Cancelling e^{-x} and using $\ln(Ae^{-x}) = \ln A - x$, we find

$$A(\ln A - x) + Ax + B = 0, \tag{3.42}$$

so that

$$A \ln A + B = 0. \tag{3.43}$$

Substituting back for B, we have a solution

$$y = e^{-x}(Ax - A \ln A), \tag{3.44}$$

where A is an arbitrary constant which can be found by imposing a boundary condition on y. ◢

Example 5 Consider

$$\frac{d^2y}{dx^2} + 2\left(\frac{dy}{dx}\right)^2 = y^2, \tag{3.45}$$

where $y = \frac{1}{4}$ when $x = 0$ and $dy/dx = \frac{1}{4}$ when $y = 0$. Putting

$$p = dy/dx, \tag{3.46}$$

we have

$$d^2y/dx^2 = p\, dp/dy. \tag{3.47}$$

Now (3.45) becomes

$$p\, dp/dy + 2p^2 = y^2. \tag{3.48}$$

This can be written

$$\frac{1}{2}\frac{d}{dy}(p^2) + 2p^2 = y^2, \tag{3.49}$$

which is a linear first-order equation in p^2. Putting $z = p^2$ in (3.49), we find

$$dz/dy + 4z = 2y^2, \tag{3.50}$$

which may be solved by multiplying by the integrating factor e^{4y}.

Accordingly,

$$\frac{d}{dy}(ze^{4y}) = 2y^2e^{4y}, \tag{3.51}$$

and hence

$$ze^{4y} = 2\int y^2e^{4y}\,dy + C, \tag{3.52}$$

where C is a constant. Performing the integration and dividing by e^{4y} gives

$$z = \tfrac{1}{2}y^2 - \tfrac{1}{4}y + \tfrac{1}{16} + Ce^{-4y}. \tag{3.53}$$

From the boundary conditions, $z = (dy/dx)^2 = \frac{1}{16}$ when $y = 0$. Hence $C = 0$ and

$$p^2 = \tfrac{1}{2}y^2 - \tfrac{1}{4}y + \tfrac{1}{16}. \tag{3.54}$$

Taking the square root,

$$p = \frac{dy}{dx} = \frac{1}{\sqrt{2}}\sqrt{[(y - \tfrac{1}{4})^2 + \tfrac{1}{16}]}, \tag{3.55}$$

the positive sign being chosen so that $p = \frac{1}{4}$ when $y = 0$. The variables may now be separated giving

$$\sqrt{2}\int \frac{dy}{\sqrt{[(y - \frac{1}{4})^2 + \frac{1}{16}]}} = x + A, \tag{3.56}$$

where A is a constant. Hence

$$\sqrt{2}\sinh^{-1}\left(\frac{y - \frac{1}{4}}{\frac{1}{4}}\right) = x + A. \tag{3.57}$$

Now $y = \frac{1}{4}$ at $x = 0$, so $A = 0$. The solution for y is therefore

$$y = \tfrac{1}{4}[1 + \sinh(x/\sqrt{2})]. \tag{3.58}$$

3.4 Bernoulli's equation

A linear equation of the type

$$dy/dx + P(x)y = Q(x) \tag{3.59}$$

may be solved, as in Example 5 above, by multiplying by the integrating factor $\exp(\int P(x)\,dx)$. The non-linear equation

$$dy/dx + P(x)y = y^nQ(x), \tag{3.60}$$

where $n \neq 0$, 1, is often called Bernoulli's equation. Using the substitution

$$1/y^{n-1} = z, \tag{3.61}$$

then

$$(1-n)y^{-n}\frac{dy}{dx} = \frac{dz}{dx}, \tag{3.62}$$

and (3.60) becomes

$$\frac{1}{1-n}\frac{dz}{dx} + P(x)z = Q(x). \tag{3.63}$$

This is now of the linear form (3.59) and hence may be solved by the integrating factor method.

Example 6 Consider

$$dy/dx + 2y = xy^3. \tag{3.64}$$

Hence

$$\frac{1}{y^3}\frac{dy}{dx} + \frac{2}{y^2} = x, \tag{3.65}$$

and therefore

$$-\frac{1}{2}\frac{d}{dx}\left(\frac{1}{y^2}\right) + \frac{2}{y^2} = x. \tag{3.66}$$

Writing $z = 1/y^2$, we have

$$dz/dx - 4z = -2x. \tag{3.67}$$

Multiplying by the integrating factor e^{-4x}, we find

$$e^{-4x}z = -2\int xe^{-4x}\,dx + C, \tag{3.68}$$

where C is a constant. Performing the integration and multiplying by e^{4x} gives

$$z = \frac{1}{y^2} = \tfrac{1}{2}x + \tfrac{1}{8} + Ce^{4x}. \tag{3.69}$$

Sometimes it is not obvious how an equation can be converted into Bernoulli form. It may be helpful to invert the equation, that is, to regard it as an equation for x in terms of y (where y is the independent variable and x is the dependent variable). To illustrate this, we give the following example.

Example 7 Consider

$$x(1-xy^2)\frac{dy}{dx}+2y=0, \tag{3.70}$$

whence

$$\frac{dy}{dx}=-\frac{2y}{x(1-xy^2)}. \tag{3.71}$$

This would seem to be difficult to solve but, by inverting, we obtain

$$\frac{dx}{dy}=-\frac{x(1-xy^2)}{2y}=-\frac{x}{2y}+\frac{x^2y}{2}, \tag{3.72}$$

and so

$$\frac{dx}{dy}+\frac{x}{2y}=\frac{yx^2}{2}. \tag{3.73}$$

We see that (3.73) is an equation of Bernoulli type when viewed as an equation for x in terms of y. Letting $z=1/x$, then

$$\frac{dz}{dy}-\frac{z}{2y}=-\frac{y}{2}, \tag{3.74}$$

for which the integrating factor is $y^{-\frac{1}{2}}$. Hence

$$zy^{-\frac{1}{2}}=-\int\frac{y}{2}y^{-\frac{1}{2}}\,dy+C, \tag{3.75}$$

giving

$$z=-\tfrac{1}{3}y^2+Cy^{\frac{1}{2}}. \tag{3.76}$$

The relationship between x and y is therefore

$$\frac{1}{x}=-\tfrac{1}{3}y^2+Cy^{\frac{1}{2}}. \tag{3.77}$$

3.5 Riccati's equation

Riccati's equation is

$$dy/dx=p(x)y^2+q(x)y+r(x), \tag{3.78}$$

where p, q and r are given functions of x. Suppose one solution of (3.78) is known, say $y=S(x)$. We then put

$$y=S(x)+1/u(x). \tag{3.79}$$

From (3.79) and the differential equation (3.78), we obtain

$$\frac{dS}{dx} - \frac{1}{u^2}\frac{du}{dx} = p(x)S^2 + q(x)S + r(x) + \frac{p(x)}{u^2} + \frac{2p(x)S}{u} + \frac{q(x)}{u}. \quad (3.80)$$

But, since S is a solution of (3.78),

$$dS/dx = p(x)S^2 + q(x)S + r(x), \quad (3.81)$$

and (3.80) becomes

$$du/dx + [2p(x)S + q(x)]u + p(x) = 0. \quad (3.82)$$

This equation is of standard first-order linear form which may be solved by the integrating factor method. Here the integrating factor is

$$v = \exp\left\{\int [2p(x)S(x) + q(x)]\, dx\right\}, \quad (3.83)$$

and consequently $u(x)$ may be obtained.

Example 8 Consider

$$\frac{dy}{dx} = y^2 + \frac{1}{x}y - \frac{1}{x} - 1. \quad (3.84)$$

Here $y = 1 = S(x)$ is a solution. Hence putting

$$y = 1 + 1/u, \quad (3.85)$$

as in (3.79), we find

$$du/dx + (2 + 1/x)u = -1. \quad (3.86)$$

For this equation the integrating factor is

$$\exp\left[\int (2 + 1/x)\, dx\right] = x\mathrm{e}^{2x}, \quad (3.87)$$

whence

$$ux\mathrm{e}^{2x} = \int -x\mathrm{e}^{2x}\, dx + C, \quad (3.88)$$

where C is a constant. Integrating and dividing by $x\mathrm{e}^{2x}$,

$$u = \frac{C}{x}\mathrm{e}^{-2x} - \frac{\mathrm{e}^{-2x}}{x}(\tfrac{1}{2}x\mathrm{e}^{2x} - \tfrac{1}{4}\mathrm{e}^{2x}). \quad (3.89)$$

From (3.85) finally

$$y = 1 + \frac{1}{\dfrac{C}{x}\mathrm{e}^{-2x} - \frac{1}{2} + \dfrac{1}{4x}}. \blacktriangleleft \quad (3.90)$$

3.6 Special forms of Riccati's equation

1.

In (3.78) we consider first the case of $p(x) = -1$ so that

$$dy/dx + y^2 = q(x)y + r(x). \tag{3.91}$$

Writing

$$y = \frac{1}{z}\frac{dz}{dx}, \tag{3.92}$$

so that

$$\frac{dy}{dx} = \frac{1}{z}\frac{d^2z}{dx^2} - \frac{1}{z^2}\left(\frac{dz}{dx}\right)^2, \tag{3.93}$$

(3.91) becomes

$$\frac{d^2z}{dx^2} - q(x)\frac{dz}{dx} - r(x)z = 0. \tag{3.94}$$

This is a linear second-order equation for z and appropriate methods may now be applied (for example, the Frobenius series method used in Chapter 2 if q and r are simple polynomials). We note that if $p(x) = +1$ in (3.78), then the substitution

$$y = -\frac{1}{z}\frac{dz}{dx} \tag{3.95}$$

will give the equation

$$\frac{d^2z}{dx^2} - q(x)\frac{dz}{dx} + r(x)z = 0. \tag{3.96}$$

This is also a linear equation.

2.

Secondly we consider the case of $q(x) = 0$ in (3.78) so that

$$dy/dx = p(x)y^2 + r(x). \tag{3.97}$$

We now make a change of the independent variable x to x', where x' is defined by

$$dx'/dx = p(x), \tag{3.98}$$

or

$$x' = \int p(x)\, dx. \tag{3.99}$$

Then

$$\frac{dy}{dx} = \frac{dy}{dx'}\frac{dx'}{dx} = p(x)\frac{dy}{dx'}. \tag{3.100}$$

Hence from (3.97) we have

$$p(x)\frac{dy}{dx'} = p(x)y^2 + r(x). \tag{3.101}$$

Equation (3.101) may be written as

$$dy/dx' = y^2 + f(x'), \tag{3.102}$$

where

$$f(x') = r[x(x')]/p[x(x')] \tag{3.103}$$

is found using the relationship (3.99) between x and x' (this may be difficult to obtain in many cases since x' is explicitly given in terms of x, but not vice-versa). Equation (3.102) may be transformed to linear form since it is a Riccati equation with $p(x) = +1$ as discussed in ***1*** above.

Example 9 Consider

$$dy/dx = 2xy^2 - 2x^3. \tag{3.104}$$

This is a case of (3.78) with $q(x) = 0$. We therefore change the variable from x to

$$x' = \int 2x\,dx = x^2, \tag{3.105}$$

following (3.99). Then

$$\frac{dy}{dx} = \frac{dy}{dx'}\frac{dx'}{dx} = 2x\frac{dy}{dx'} \tag{3.106}$$

and substituting this into (3.104) and cancelling $2x$ gives

$$dy/dx' = y^2 - x^2 = y^2 - x'. \tag{3.107}$$

This is now a Riccati equation for y in terms of x' with $p(x')$ in (3.78) equal to $+1$. Following (3.95), we put

$$y = -\frac{1}{z}\frac{dz}{dx'} \tag{3.108}$$

and obtain from (3.107)

$$\frac{d^2z}{dx'^2} = x'z. \tag{3.109}$$

This is Airy's equation discussed in Section 2.9 and hence a solution for z in terms of x' may be found in terms of Airy functions. The solution for y is then obtained from (3.108) and finally substituting x^2 for x' from (3.105). ◢

3.7 The Lane–Emden equation

The form of this equation is

$$\frac{d^2y}{dx^2}+\frac{2}{x}\frac{dy}{dx}+y^{\alpha}=0, \tag{3.110}$$

where α is a constant. Solutions which satisfy the boundary conditions

$$y=1, \quad dy/dx=0 \quad \text{at} \quad x=0, \tag{3.111}$$

are called Lane–Emden functions. Exact solutions are known when $\alpha=0$, 1 and 5. In the cases $\alpha=0$ and $\alpha=1$, the equation is linear and the solutions are easily found. When $\alpha=5$, let

$$x=\mathrm{e}^{-t}, \quad y=\frac{1}{\sqrt{2}}\mathrm{e}^{t/2}u(t). \tag{3.112}$$

Then after some algebra we find

$$4\,d^2u/dt^2-u+u^5=0. \tag{3.113}$$

Now writing $v=du/dt$, so that $d^2u/dt^2=v\,dv/du$, (3.113) becomes

$$4v\,dv/du-u+u^5=0. \tag{3.114}$$

This equation may be integrated with respect to u to give

$$2v^2-\tfrac{1}{2}u^2+\tfrac{1}{6}u^6=C, \tag{3.115}$$

where C is a constant. From the boundary conditions (3.111) and the transformations (3.112), we see that $x=0$ corresponds to $t\to\infty$ and hence u and du/dt must tend to zero as $t\to\infty$. Hence $u=v=0$ as $t\to\infty$ and so, from (3.115), $C=0$ giving

$$v^2=\tfrac{1}{4}u^2-\tfrac{1}{12}u^6, \tag{3.116}$$

or

$$v=du/dt=\tfrac{1}{2}u\sqrt{(1-\tfrac{1}{3}u^4)}. \tag{3.117}$$

Separating the variables and integrating, we have

$$2\int\frac{du}{u\sqrt{(1-\frac{1}{3}u^4)}}=\int dt=t+A, \tag{3.118}$$

where A is a constant. To evaluate the integral on the left-hand side, we put $u^4=3\cos^2\theta$ so that

$$t+A=-\int\sec\theta\,\mathrm{d}\theta=-\ln(\sec\theta+\tan\theta). \tag{3.119}$$

Taking the exponential of both sides yields, with $B = e^{-A}$,

$$Be^{-t} = \sec\theta + \tan\theta. \tag{3.120}$$

Using $\cos\theta = u^2/\sqrt{3}$ and $x = e^{-t}$, we find

$$Bx = \frac{\sqrt{3}}{u^2} + \sqrt{\left(\frac{3}{u^4} - 1\right)}, \tag{3.121}$$

the solution of which is

$$u = \sqrt{\left(\frac{2\sqrt{3}Bx}{1 + B^2x^2}\right)}. \tag{3.122}$$

From (3.112), the solution for y is

$$y = \sqrt{\left(\frac{\sqrt{3}B}{1 + B^2x^2}\right)}. \tag{3.123}$$

Applying the boundary condition (3.111) that $y = 1$ when $x = 0$ gives $B = 1/\sqrt{3}$ and, finally,

$$y = 1/\sqrt{(1 + \tfrac{1}{3}x^2)}. \tag{3.124}$$

3.8 The non-linear pendulum

The non-linear equation

$$d^2\theta/dt^2 + \omega^2 \sin\theta = 0, \tag{3.125}$$

where ω is a constant, is called the non-linear pendulum equation and may be solved in terms of a class of integrals known as elliptic integrals. Multiplying (3.125) by $2\,d\theta/dt$ and integrating, we have

$$(d\theta/dt)^2 = 2\omega^2\cos\theta + C, \tag{3.126}$$

where C is a constant of integration. Assume, for example, that $d\theta/dt = 0$ when $\theta = \alpha$. Then $C = -2\omega^2\cos\alpha$ and

$$(d\theta/dt)^2 = 2\omega^2(\cos\theta - \cos\alpha). \tag{3.127}$$

Taking the square root, separating the variables and integrating, we have

$$t = \frac{1}{\omega\sqrt{2}}\int_0^\theta \frac{d\theta}{\sqrt{(\cos\theta - \cos\alpha)}}, \tag{3.128}$$

assuming $\theta = 0$ when $t = 0$. If $t = T$ at $\theta = \theta_1$, then

$$T = \frac{1}{\omega\sqrt{2}}\int_0^{\theta_1} \frac{d\theta}{\sqrt{(\cos\theta - \cos\alpha)}}. \tag{3.129}$$

Using

$$\cos\theta = 1 - 2\sin^2(\theta/2), \quad \cos\alpha = 1 - 2\sin^2(\alpha/2), \tag{3.130}$$

(3.129) becomes

$$T = \frac{1}{2\omega}\int_0^{\theta_1} \frac{d\theta}{\sqrt{[\sin^2(\alpha/2) - \sin^2(\theta/2)]}}. \tag{3.131}$$

Now writing

$$\sin(\theta/2) = k\sin\phi, \tag{3.132}$$

where $k = \sin(\alpha/2)$, then

$$\omega T = \int_0^{\phi_1} \frac{d\phi}{\sqrt{(1 - k^2\sin^2\phi)}}, \tag{3.133}$$

in which $k\sin\phi_1 = \sin(\theta_1/2)$. The integral in (3.133) is referred to as an elliptic integral of the first kind and is usually denoted by

$$F(k, \phi_1) = \int_0^{\phi_1} \frac{d\phi}{\sqrt{(1 - k^2\sin^2\phi)}}, \tag{3.134}$$

with $0 \leq k \leq 1$. Similarly, the elliptic integral of the second kind is defined by

$$E(k, \phi_1) = \int_0^{\phi_1} \sqrt{(1 - k^2\sin^2\phi)}\, d\phi, \tag{3.135}$$

where, as before, $0 \leq k \leq 1$. In both cases the elliptic integrals are said to be complete if $\phi_1 = \pi/2$ and are then denoted by $F(k)$ and $E(k)$, respectively. Graphs of $F(k, \phi_1)$ and $E(k, \phi_1)$ are shown in Figures 3.1(a) and 3.1(b). We see that (3.133) gives T explicitly in terms of ϕ_1 but not vice-versa.

We mention here that other integrals may be expressed in terms of the above elliptic integrals and hence evaluated by consulting tables of values (see reference on page 22).

Example 10 Evaluate

$$I_1 = \int_0^{\pi/6} \frac{d\phi}{\sqrt{(1 - 4\sin^2\phi)}}. \tag{3.136}$$

Putting $4\sin^2\phi = \sin^2\theta$, we obtain

$$I_1 = \tfrac{1}{2}\int_0^{\pi/2} \frac{d\theta}{\sqrt{(1 - \frac{1}{4}\sin^2\theta)}} = \tfrac{1}{2}F\left(\tfrac{1}{2}, \frac{\pi}{2}\right) = \tfrac{1}{2}F(\tfrac{1}{2}). \quad ◢ \tag{3.137}$$

Example 11 Evaluate

$$I_2 = \int_0^{\frac{1}{2}} \frac{dx}{\sqrt{(3 - 4x^2 + x^4)}}. \tag{3.138}$$

Putting $x = \sin\phi$, we obtain

$$I_2 = \int_0^{\pi/6} \frac{\cos\phi \, d\phi}{\sqrt{(3 - 4\sin^2\phi + \sin^4\phi)}}. \tag{3.139}$$

However, $3 - 4\sin^2\phi + \sin^4\phi = (3 - \sin^2\phi)(1 - \sin^2\phi) = (3 - \sin^2\phi)\cos^2\phi$. Hence

$$I_2 = \frac{1}{\sqrt{3}} \int_0^{\pi/6} \frac{d\phi}{\sqrt{(1 - \frac{1}{3}\sin^2\phi)}} = \frac{1}{\sqrt{3}} F\left(\frac{1}{\sqrt{3}}, \frac{\pi}{6}\right). \tag{3.140}$$

3.9 Duffing's equation

In its simplest form Duffing's equation is

$$d^2y/dx^2 + ay + by^3 = 0, \tag{3.141}$$

where a and b are constants. We proceed, as in the case of the

Figure 3.1

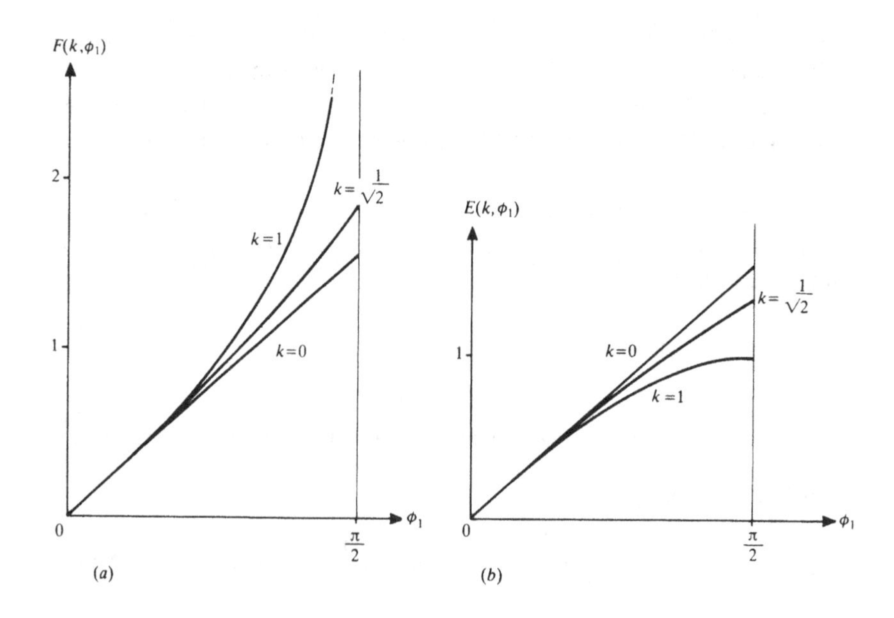

non-linear pendulum equation, by multiplying (3.141) by $2\,dy/dx$. Then

$$2\frac{dy}{dx}\frac{d^2y}{dx^2}=2(-ay-by^3)\frac{dy}{dx}. \tag{3.142}$$

This equation may be integrated directly to give

$$\left(\frac{dy}{dx}\right)^2=C-ay^2-\tfrac{1}{2}by^4, \tag{3.143}$$

where C is a constant of integration. Taking the square root of (3.143) and separating the variables, we find

$$\int\frac{dy}{\sqrt{(C-ay^2-\frac{1}{2}by^4)}}=x+B, \tag{3.144}$$

where B is another constant. For particular values of a, b and C, the left-hand side can be expressed in terms of an elliptic integral.

Example 12 If

$$d^2y/dx^2=-\tfrac{3}{2}y+y^3, \tag{3.145}$$

with boundary conditions $y=0$ and $dy/dx=1$ when $x=0$, find the x-value for which $y=1$.

Proceeding as above, we have

$$2\frac{dy}{dx}\frac{d^2y}{dx^2}=2(-\tfrac{3}{2}y+y^3)\frac{dy}{dx}, \tag{3.146}$$

which on integration gives

$$(dy/dx)^2=C-\tfrac{3}{2}y^2+\tfrac{1}{2}y^4, \tag{3.147}$$

where C is a constant. Since $y=0$ when $dy/dx=1$, we find $C=1$ and

$$dy/dx=\sqrt{(1-\tfrac{3}{2}y^2+\tfrac{1}{2}y^4)}. \tag{3.148}$$

Separating the variables and using $y=0$ when $x=0$,

$$\int_0^y\frac{dy}{\sqrt{(1-\frac{3}{2}y^2+\frac{1}{2}y^4)}}=x. \tag{3.149}$$

The value of x, x_0 say, for which $y=1$ is

$$x_0=\int_0^1\frac{dy}{\sqrt{(1-\frac{3}{2}y^2+\frac{1}{2}y^4)}}. \tag{3.150}$$

Now putting $y=\sin\phi$, we have $dy=\cos\phi\,d\phi$ and

$$1-\tfrac{3}{2}y^2+\tfrac{1}{2}y^4=\tfrac{1}{2}(2-3\sin^2\phi+\sin^4\phi)=(1-\tfrac{1}{2}\sin^2\phi)\cos^2\phi. \tag{3.151}$$

Hence

$$x_0 = \int_0^{\pi/2} \frac{d\phi}{\sqrt{(1 - \frac{1}{2}\sin^2\phi)}} = F\left(\frac{1}{\sqrt{2}}\right), \tag{3.152}$$

using the definition of the complete elliptic integral of the first kind. ◢

In general, it is extremely difficult to obtain analytic (closed form) solutions to Duffing's equation. However, a technique known as the phase-plane method is useful in finding the nature of the solutions of (3.141) for any a and b values and given initial conditions. We define

$$dy/dx = w, \tag{3.153}$$

so that (3.141) becomes

$$dw/dx = -ay - by^3. \tag{3.154}$$

The equations (3.153) and (3.154) are now a particular case of the general system

$$\left.\begin{aligned} dy/dx &= P(y, w), \\ dw/dx &= Q(y, w), \end{aligned}\right\} \tag{3.155}$$

where, for Duffing's equation,

$$P(y, w) = w, \quad Q(y, w) = -ay - by^3. \tag{3.156}$$

It is for a system of the form (3.155) that the phase-plane method is appropriate. By eliminating x from (3.155), we have

$$\frac{dy}{dw} = \frac{P(y, w)}{Q(y, w)}. \tag{3.157}$$

In particular, for the Duffing equation,

$$\frac{dy}{dw} = -\frac{w}{ay + by^3}. \tag{3.158}$$

Separating the variables and integrating, we find the solution of (3.158) to be

$$\tfrac{1}{2}ay^2 + \tfrac{1}{4}by^4 + \tfrac{1}{2}w^2 = C, \tag{3.159}$$

where C is a constant determined by the initial conditions. The $(y, w) \equiv (y, dy/dx)$ plane is called the phase-plane and the nature of the solution (3.159) of (3.158), as represented by the curves of y and w in the phase-plane for different values of C, yields some information about the solution of Duffing's equation, but does not give the solution itself. For example, it can be shown that the existence of simple (non-intersecting) closed curves in the phase-plane implies the existence of periodic solutions of the original equation. Specifically, Duffing's equation will have periodic solutions if there exist simple closed (y, w) curves described by (3.159). To illustrate this, we consider the following values of a and b:

1. $a = 1,\ b = 2$

In this case (3.159) becomes

$$y^4 + y^2 + w^2 = 2C. \tag{3.160}$$

The graph of this family of curves is shown in Figure 3.2. From (3.160), we see that it is not possible to have $C < 0$ for any initial conditions. For any value of $C \geqslant 0$, the curve is closed, which indicates that only periodic solutions of the equation exist for these values of a and b.

2. $a = 1,\ b = -2$

In this case (3.159) becomes

$$y^4 - y^2 - w^2 = -2C. \tag{3.161}$$

Figure 3.2

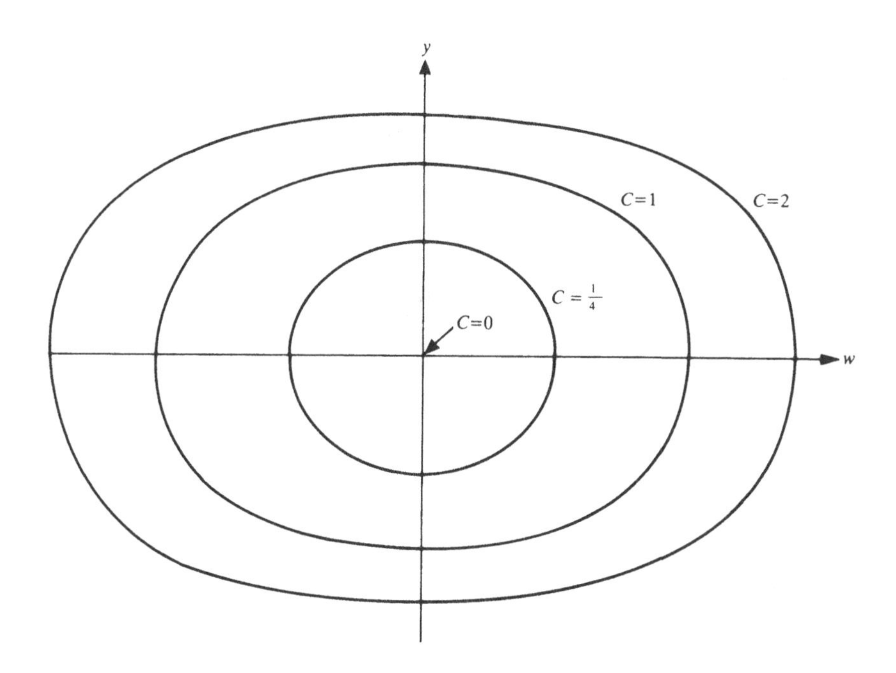

The graph of this family of curves is shown in Figure 3.3. If $0 \leq C < \frac{1}{8}$ both open and closed curves exist in the phase-plane, whereas if $C < 0$ or $C \geq \frac{1}{8}$ only open curves exist. Hence, *periodic* solutions of the equation exist only if the initial conditions are such that $0 \leq C < \frac{1}{8}$, *and* at $t = 0$ the point lies *within* the shaded region of Figure 3.3. An example of this is shown for $C = \frac{1}{16}$. The phase-plane method is also of importance in problems of stability of solutions of non-linear systems of the form (3.155). The technique requires knowledge of the fixed or critical points (y_0, w_0) of (3.155) defined by

$$P(y_0, w_0) = Q(y_0, w_0) = 0. \tag{3.162}$$

The stability problem is not central to our discussion of obtaining solutions of non-linear equations and, accordingly, we shall not pursue it here.

For the Duffing equation, the differential relation (3.158) defining the curves in the phase-plane is simple to integrate since the variables may be separated. This may not be the case for other non-linear

Figure 3.3

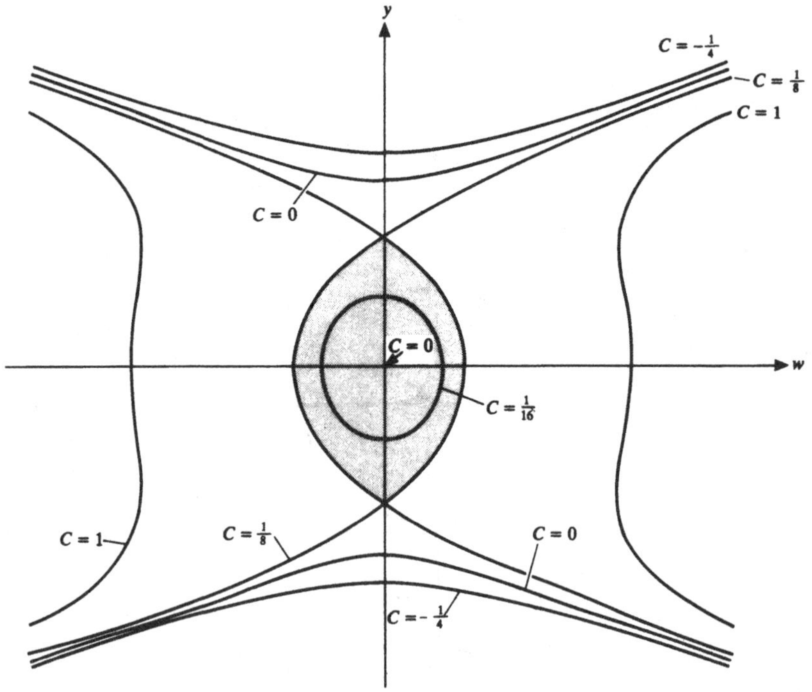

equations. For example, the van der Pol equation

$$\frac{d^2y}{dx^2} + \mu(y^2 - 1)\frac{dy}{dx} + y = 0, \tag{3.163}$$

where μ is a constant, has curves in the phase-plane defined by

$$\frac{dy}{dw} = -\frac{w}{\mu(y^2 - 1)w + y}. \tag{3.164}$$

This cannot be integrated by separation of variables but solution curves can be sketched using, for example, the method of isoclines†.

For a more detailed account of analytic techniques for studying non-linear differential equations, including stability problems and chaotic behaviour, the reader is referred to a standard text‡. In general, however, non-linear equations require numerical methods for their solution.

Problems 3

1. Obtain a solution of the equation

$$y'y'' + (y')^2 = 2,$$

given $y(0) = 0$ and $y'(0) = \sqrt{2}$.

2. By writing $dy/dx = p$, solve

$$y'' - \frac{1}{y}(y')^2 = \tfrac{1}{2}y^2,$$

given $y = dy/dx = 1$ when $x = 1$.

3. By writing $xy' - y = v(x)$, solve

$$x^3y'' = (xy' - y)^2.$$

4. By writing $e^y = z$, solve

$$y' + \frac{1}{\ln x}e^y = \frac{1}{x \ln x}.$$

5. Show that the equation

$$y'' = xy + 2y^3 + \tfrac{1}{2}$$

† E. Kreysig, *Advanced Engineering Mathematics* (Wiley, New York, 1988) Section 1.10.

‡ D. W. Jordan and P. Smith, *Nonlinear Ordinary Differential Equations*, (O.U.P, Oxford, 1987)

has a first integral $y' = \frac{1}{2}x + y^2$. By finding the general solution of this Riccati equation in terms of Airy functions, obtain a solution of the original equation containing one arbitrary constant.

6. Verify that

$$y = \sqrt{\left(u^2 + \frac{C}{W^2} v^2\right)},$$

where C is a constant, satisfies Pinney's equation

$$y'' + p(x)y = C/y^3,$$

where u and v are independent solutions of

$$z'' + p(x)z = 0$$

and $W = uv' - vu'$. (For equations with no first derivative term dz/dx, W, the Wronskian, is a constant). Hence show that

$$y = \sqrt{(\sin^2 x + 2\cos^2 x)}$$

is a solution of

$$y'' + y = 2/y^3.$$

7. Show that

(i) $$\int_{2/\sqrt{3}}^{\infty} \frac{du}{\sqrt{(2u^4 - 3u^2 + 1)}} = \frac{1}{\sqrt{2}} F\left(\frac{1}{\sqrt{2}}, \frac{\pi}{3}\right),$$

(ii) $$\int_0^{\infty} \sqrt{\left[\frac{9 + 8x^2}{(1 + x^2)^3}\right]}\, dx = 3E(\tfrac{1}{3}),$$

where F and E are elliptic integrals of the first and second kinds, respectively.

8. Show that the inhomogeneous Duffing equation

$$y'' + ay + by^3 = A\cos(3\omega t)$$

has an exact solution

$$y = \left(\frac{4A}{b}\right)^{\frac{1}{3}} \cos(\omega t),$$

provided

$$\omega^2 = a + 3(A^2 b/4)^{\frac{1}{3}}.$$

4
Approximate solutions of ordinary differential equations

4.1 Power series

We begin by recalling briefly some elementary ideas. Linear equations of the form

$$\frac{dy}{dx}+f(x)y=0 \tag{4.1}$$

or

$$\frac{d^2y}{dx^2}+p(x)\frac{dy}{dx}+q(x)y=0 \tag{4.2}$$

can be solved in many cases by substituting a Taylor series expansion about some fixed point x_0 of the form

$$y=\sum_{n=0}^{\infty} a_n(x-x_0)^n. \tag{4.3}$$

Provided $f(x)$, $p(x)$ and $q(x)$ are also expressible as Taylor series about x_0, then the coefficients of like powers of $x-x_0$ may be equated in the differential equation and, by solving for the coefficients a_n, the solutions of the equation may be obtained. We shall mostly be concerned here with the case $x_0=0$ for which the series (4.3) becomes a MacLaurin series.

Example 1 Consider

$$dy/dx=2e^{-x}-y, \tag{4.4}$$

given that $y(0)=0$.

Writing $y = \sum_{n=0}^{\infty} a_n x^n$, substituting this into (4.4), and expanding e^{-x} as a power series, gives

$$\sum_{n=0}^{\infty} a_n n x^{n-1} = 2\left(1 - x + \frac{x^2}{2!} - \frac{x^3}{3!} + \ldots\right) - \sum_{n=0}^{\infty} a_n x^n. \tag{4.5}$$

The lowest power of x on each side is x^0, and equating coefficients of powers of $x^0, x^1, x^2, x^3, \ldots$ gives, respectively,

$$a_1 = 2 - a_0, \tag{4.6}$$

$$2a_2 = -2 - a_1, \tag{4.7}$$

$$3a_3 = \frac{2}{2!} - a_2, \tag{4.8}$$

$$4a_4 = -\frac{2}{3!} - a_3, \tag{4.9}$$

and so on. Hence, since $y(0) = 0$, $a_0 = 0$ and therefore

$$a_1 = 2, \quad a_2 = -2, \quad a_3 = 1, \quad a_4 = -\tfrac{1}{3}, \tag{4.10}$$

giving

$$y(x) = 2x - 2x^2 + x^3 - \tfrac{1}{3}x^4 + \ldots. \tag{4.11}$$

The coefficient of x^n for $n \geqslant 1$ is easily seen to be

$$a_n = \frac{2(-1)^{n-1}}{(n-1)!}. \tag{4.12}$$

Hence, applying the ratio test for convergence, we find

$$|a_{n+1}/a_n| = 1/n. \tag{4.13}$$

Consequently the series (4.11) converges for all values of x since the ratio of the $(n+1)$th term to the nth is x/n, which tends to zero for any finite x as $n \to \infty$. The exact solution of (4.4) is easily found by the integrating factor method to be $y(x) = 2xe^{-x}$, of which (4.11) is the MacLaurin expansion.

This method is equivalent to assuming a MacLaurin series

$$y(x) = y(0) + xy'(0) + \frac{x^2}{2!}y''(0) + \ldots \tag{4.14}$$

and obtaining the derivatives of y by successive differentiation of the equation. From (4.4),

$$d^2y/dx^2 = -2e^{-x} - dy/dx \tag{4.15}$$

and, in general,

$$\frac{d^n y}{dx^n} = (-1)^{n-1} 2e^{-x} - \frac{d^{n-1}y}{dx^{n-1}}. \tag{4.16}$$

Hence, at $x = 0$,

$$y^{(n)}(0) = 2(-1)^{n-1} - y^{(n-1)}(0). \tag{4.17}$$

Since $y(0) = 0$, we find from (4.17) that $y'(0) = 2$, $y''(0) = -4$ and so on. On substituting these into (4.12) we again obtain the solution (4.11). ◢

Although the power series method is readily justified in the case of linear equations, it is more difficult to justify its use for non-linear equations of the type $dy/dx = f(x, y)$. Proofs of convergence and of the existence of solutions are complicated in the case of non-linear equations, and power series methods should be used with caution. We illustrate this with an example.

Example 2 Consider

$$dy/dx = x - y^2, \tag{4.18}$$

given that $y(0) = 0$.

Writing, as before, $y = \sum_{n=0}^{\infty} a_n x^n$ and substituting in (4.18) gives

$$a_1 + 2a_2 x + 3a_3 x^2 + \ldots = x - (a_0 + a_1 x + \ldots)^2. \tag{4.19}$$

Comparing coefficients of like powers of x gives

$$a_1 = -a_0^2, \tag{4.20}$$

$$2a_2 = 1 - 2a_0 a_1, \tag{4.21}$$

$$3a_3 = -2a_0 a_2 - a_1^2, \tag{4.22}$$

$$4a_4 = -2a_1 a_2 - 2a_0 a_3, \tag{4.23}$$

$$5a_5 = -2a_1 a_3 - a_2^2 - 2a_0 a_4, \tag{4.24}$$

and so on. Since $y(0) = 0$, we have from the original series $a_0 = 0$. Hence, from (4.20)–(4.24),

$$a_1 = 0, \quad a_2 = \tfrac{1}{2}, \quad a_3 = 0, \quad a_4 = 0, \quad a_5 = -\tfrac{1}{20}, \tag{4.25}$$

and so on. We have finally

$$y(x) = \tfrac{1}{2}x^2 - \tfrac{1}{20}x^5 + \ldots. \tag{4.26}$$

Although a series solution can be generated in this way, it is difficult to find an expression for the general term and hence to prove convergence. ◢

4.2 Frobenius series

In the case of the second-order equation

$$\frac{d^2y}{dx^2}+p(x)\frac{dy}{dx}+q(x)y=0, \tag{4.27}$$

where $p(x)$ and $q(x)$ are given functions, a power series solution does not necessarily exist. However, provided $p(x)$ and $q(x)$ are differentiable and single-valued at a point x_0 (that is, they are regular at x_0), then x_0 is called an ordinary point of the equation and a Taylor series centred at x_0 will provide the solutions. If, however, x_0 is not an ordinary point but nevertheless

$$\lim_{x\to x_0}(x-x_0)p(x) \tag{4.28}$$

and

$$\lim_{x\to x_0}(x-x_0)^2q(x) \tag{4.29}$$

are finite at x_0, then x_0 is called a regular singular point of the equation and near this point it is possible to find at least one solution of the form

$$y(x)=\sum_{n=0}^{\infty}a_n(x-x_0)^{m+n}, \tag{4.30}$$

where m is some number. If (4.28) and (4.29) are not finite, then x_0 is an irregular singular point. The series (4.30) is known as a Frobenius series and has been used in Section 2.5 to develop the solution of the Bessel equation. In applying this method, certain special cases arise depending on the two values m_1 and m_2 of m. If these values are identical or differ by an integer, then only one solution can be found by this method. This was the case with the Bessel equation of integer order n for which m_1-m_2 is an integer $2n$ and (see (2.92)) $J_{-n}(x)=(-1)^nJ_n(x)$.

We now give a few examples of equations, taking $x_0=0$ as the point of expansion.

(i)

$$\frac{d^2y}{dx^2}+x\frac{dy}{dx}+2y=0. \tag{4.31}$$

For this equation $x = 0$ is an ordinary point and power series solutions exist.

(ii)
$$\frac{d^2y}{dx^2} + \frac{3}{x}\frac{dy}{dx} + \frac{1}{x^2}y = 0. \tag{4.32}$$

Here $x = 0$ is a regular singular point since $\lim_{x\to 0} x(3/x) = 3$ and $\lim_{x\to 0} x^2(1/x^2) = 1$ are finite. At least one Frobenius series solution exists.

(iii)
$$\frac{d^2y}{dx^2} + \frac{1}{x^2}\frac{dy}{dx} + xy = 0. \tag{4.33}$$

In this case $x = 0$ is an irregular singular point since $\lim_{x\to 0} x(1/x^2)$ does not exist.

We note that $x = 0$ is a regular singular point of the Bessel equation (2.61) since $\lim_{x\to 0} xp(x) = \lim_{x\to 0} x(1/x) = 1$ and $\lim_{x\to 0} x^2q(x) = \lim_{x\to 0} (x^2 - v^2) = -v^2$ are finite. The application of the Frobenius series method follows that of Section 2.5 and we shall not illustrate the method further.

4.3 Picard iterative method

Consider the standard first-order equation

$$dy/dx = f(x, y), \tag{4.34}$$

with $y(x_0) = y_0$ given. Equation (4.34) has the formal solution

$$y(x) = A + \int_{x_0}^{x} f(x, y)\, dx, \tag{4.35}$$

where, since $y(x_0) = y_0$, $A = y_0$.

A sequence of functions $y_1, y_2, \ldots, y_n$ is generated as follows: first we insert y_0 into the right-hand side of (4.34) to get

$$dy_1/dx = f(x, y_0). \tag{4.36}$$

Solving for y_1 and inserting this into (4.34) gives

$$dy_2/dx = f(x, y_1), \tag{4.37}$$

and so on. Hence we form the sequence by iterating the relation

$$y_{n+1}(x) = y_0 + \int_{x_0}^{x} f(x, y_n)\, dx, \tag{4.38}$$

where $n = 0, 1, 2, \ldots,$ from (4.35). It can be proved that this sequence tends to a solution of (4.34) on some interval of x, except in particular circumstances (see Example 6).

Example 3 Consider

$$\frac{dy}{dx}=\frac{x^2+y^2}{x}, \tag{4.39}$$

given $y(0)=0$. (This is a Riccati equation – see Section 3.5.) Then

$$y_{n+1}(x)=0+\int_0^x \frac{x^2+y_n^2(x)}{x}\,dx. \tag{4.40}$$

Hence

$$y_1(x)=\int_0^x \frac{x^2+0}{x}\,dx=\frac{x^2}{2}, \tag{4.41}$$

$$y_2(x)=\int_0^x \frac{x^2+(x^2/2)^2}{x}\,dx=\frac{x^2}{2}+\frac{x^4}{16}, \tag{4.42}$$

$$y_3(x)=\int_0^x \frac{x^2+(x^2/2+x^4/16)^2}{x}\,dx=\frac{x^2}{2}+\frac{x^4}{16}+\frac{x^6}{96}+\frac{x^8}{2048}. \tag{4.43}$$

Each of these approximations is a finite series. This sequence approaches the exact solution as $n\to\infty$. ◢

Example 4 Consider

$$dy/dx=x+\sin y, \tag{4.44}$$

with $y(0)=\pi/2$. Then

$$y_{n+1}(x)=\frac{\pi}{2}+\int_0^x (x+\sin y_n)\,dx. \tag{4.45}$$

Hence

$$y_1(x)=\frac{\pi}{2}+\int_0^x (x+1)\,dx=\frac{\pi}{2}+\frac{x^2}{2}+x. \tag{4.46}$$

However,

$$y_2(x)=\frac{\pi}{2}+\int_0^x \left[x+\sin\left(\frac{\pi}{2}+\frac{x^2}{2}+x\right)\right]dx \tag{4.47}$$

cannot be integrated analytically, and so the next function in the sequence cannot be found except, perhaps, for sufficiently small values of x for which the sine term in (4.47) may be expanded. ◢

The Picard method of successive approximation may be used on a system of equations. Suppose we want to solve for $y(x)$ and $z(x)$, where

$$dy/dx=f(x,y,z), \quad dz/dx=g(x,y,z), \tag{4.48}$$

given, say, $y(0) = y_0$ and $z(0) = z_0$. Then the iterated sequence is given by

$$y_{n+1} = y_0 + \int_0^x f(x, y_n, z_n)\, dx, \quad z_{n+1} = z_0 + \int_0^x g(x, y_n, z_n)\, dx. \tag{4.49}$$

Example 5 Consider

$$dy/dx = x + z, \quad dz/dx = x - y^2, \tag{4.50}$$

where $y(0) = 2,\ z(0) = 1$. Then

$$y_1(x) = 2 + \int_0^x (x + 1)\, dx = 2 + x + \tfrac{1}{2}x^2, \tag{4.51}$$

$$z_1(x) = 1 + \int_0^x (x - 2^2)\, dx = 1 - 4x + \tfrac{1}{2}x^2. \tag{4.52}$$

Similarly

$$y_2(x) = 2 + \int_0^x (x + 1 - 4x + \tfrac{1}{2}x^2)\, dx = 2 + x - \tfrac{3}{2}x^2 + \tfrac{1}{6}x^3, \tag{4.53}$$

$$z_2(x) = 1 + \int_0^x (x - [2 + x + \tfrac{1}{2}x^2]^2)\, dx = 1 - 4x - \tfrac{3}{2}x^2 - x^3 - \tfrac{1}{4}x^4 - \tfrac{1}{20}x^5, \tag{4.54}$$

and so on. ◢

The Picard method does not always provide a non-trivial solution as we demonstrate in the next example.

Example 6 Consider

$$dy/dx = \sqrt{y}, \tag{4.55}$$

given $y(0) = 0$. Exact solutions of this equation are $y = 0$ and $y = \frac{1}{4}x^2$. Applying the Picard method, we find

$$y_{n+1}(x) = y(0) + \int_0^x \sqrt{[y_n(x)]}\, dx. \tag{4.56}$$

Hence

$$y_1(x) = 0 + \int_0^x 0\, dx = 0, \tag{4.57}$$

$$y_2(x) = 0 + \int_0^x 0\, dx = 0, \tag{4.58}$$

and so on. For this equation, the Picard method only generates the trivial solution $y = 0$. ◢

The failure in Example 6 to produce the non-trivial solution is related to the Lipschitz condition which states that for the general equation (4.34), $|\partial f/\partial y|$ must be *finite* in the range of integration for a non-trivial solution to be generated. In Example 6, $f(x, y) = \sqrt{y}$ and hence $|\partial f/\partial y| = 1/2\sqrt{y}$ which tends to *infinity* as $y \to 0$.

4.4 Perturbation series

An important technique for solving non-linear ordinary differential equations is the perturbation method. This method requires that the equation is non-linear in virtue of a small parameter, say ϵ. By expanding the dependent variable $y(t)$, say, as a power series in ϵ we generate a solution to any desired order in ϵ. Consider the following typical example.

Example 7 Duffing's equation (see Section 3.9) has the form

$$d^2y/dt^2 + ay + by^3 = 0. \tag{4.59}$$

Suppose we take the case where $a = 1$ and $b = \epsilon$, where ϵ is a small parameter. Suppose also that $y(0) = 1$ and $y'(0) = 0$. We expand y as a series in ϵ, as follows:

$$y(t) = y_0(t) + \epsilon y_1(t) + \epsilon^2 y_2(t) + \dots. \tag{4.60}$$

Substituting this series into (4.59) gives

$$\left(\frac{d^2y_0}{dt^2} + \epsilon\frac{d^2y_1}{dt^2} + \epsilon^2\frac{d^2y_2}{dt^2} + \dots\right)$$

$$+ (y_0 + \epsilon y_1 + \epsilon^2 y_2 + \dots) + \epsilon(y_0 + \epsilon y_1 + \dots)^3 = 0. \tag{4.61}$$

Equating to zero the coefficients of successive powers of ϵ gives

$$d^2y_0/dt^2 + y_0 = 0, \tag{4.62}$$

$$d^2y_1/dt^2 + y_1 + y_0^3 = 0, \tag{4.63}$$

$$d^2y_2/dt^2 + y_2 + 3y_0^2 y_1 = 0, \tag{4.64}$$

and so on. Now, given $y(0) = 1$ and $y'(0) = 0$, we have from (4.60) on equating powers of ϵ, $y_0(0) = 1$, $y_0'(0) = 0$,

$$y_1(0) = y_2(0) = \dots = 0 \tag{4.65}$$

and

$$y_1'(0) = y_2'(0) = \dots = 0. \tag{4.66}$$

Hence (4.62) gives

$$y_0(t) = \cos t, \tag{4.67}$$

whilst (4.63) gives

$$d^2y_1/dt^2 + y_1 + \cos^3 t = 0. \tag{4.68}$$

This is a linear equation with constant coefficients which can be solved by elementary methods. Writing $\cos^3 t = \frac{1}{4}(\cos 3t + 3\cos t)$, we find using (4.65) and (4.66) that

$$y_1(t) = -\tfrac{1}{32}(\cos t - \cos 3t) - \tfrac{3}{8} t \sin t. \tag{4.69}$$

This may be inserted into (4.64), together with (4.67), to give $y_2(t)$ and so on, giving a solution of the form (4.60) to whatever order in ϵ is required. ◢

The previous example is only meant to illustrate the general approach. The method may also be applied to the van der Pol equation (see (3.163)) where μ is a small parameter. To proceed further would take us too far into technical details (see reference ‡ on page 73).

4.5 Normal form

In the following sections we shall be dealing with approximation methods for solving second-order linear differential equations in normal form for which there is no first derivative term. This is no real restriction since any second-order linear differential equation can be cast into normal form as we now show.

Consider the equation

$$\frac{d^2y}{dx^2} + p(x)\frac{dy}{dx} + q(x)y = 0. \tag{4.70}$$

Now let $y = u(x)v(x)$. Then

$$u''v + 2u'v' + uv'' + p(uv' + vu') + quv = 0. \tag{4.71}$$

We now choose u so that the first derivative term in v' vanishes by putting

$$2u' + pu = 0. \tag{4.72}$$

Hence

$$u'/u = -\tfrac{1}{2}p, \tag{4.73}$$

giving

$$u = A \exp\left\{-\tfrac{1}{2}\int^x p(x)\, dx\right\}. \tag{4.74}$$

Here A is an arbitrary constant of integration which may be taken equal to unity without loss of generality since u occurs in every term of (4.71). Inserting u' and u'', obtained from (4.74), into (4.71), we find

$$\frac{d^2v}{dx^2}+\left(q-\frac{1}{2}\frac{dp}{dx}-\tfrac{1}{4}p^2\right)v=0. \tag{4.75}$$

Hence the solution of (4.70) may be obtained from (4.74) and the solution of (4.75), which is an equation in normal form.

Example 8 Reduce

$$\frac{d^2y}{dx^2}+x\frac{dy}{dx}+\tfrac{1}{4}x^2y=0 \tag{4.76}$$

to normal form and hence obtain its general solution.

Proceeding as above with $p(x)=x$ and $q(x)=\tfrac{1}{4}x^2$, we find from (4.74)

$$u=\exp\left\{-\tfrac{1}{2}\int^x x\,dx\right\}=\mathrm{e}^{-\frac{1}{4}x^2}, \tag{4.77}$$

and from (4.75),

$$d^2v/dx^2+(\tfrac{1}{4}x^2-\tfrac{1}{2}-\tfrac{1}{4}x^2)v=0 \tag{4.78}$$

or

$$d^2v/dx^2-\tfrac{1}{2}v=0. \tag{4.79}$$

Hence

$$v=A\mathrm{e}^{x/\sqrt{2}}+B\mathrm{e}^{-x/\sqrt{2}} \tag{4.80}$$

and finally therefore

$$y=uv=\mathrm{e}^{-\frac{1}{4}x^2}(A\mathrm{e}^{x/\sqrt{2}}+B\mathrm{e}^{-x/\sqrt{2}}), \tag{4.81}$$

where A and B are arbitrary constants. ◢

4.6 The W.K.B. (Wentzel–Kramers–Brillouin) approximation

This is a method of obtaining an approximate solution to any linear second-order differential equation in normal form when the second-order derivative term is multiplied by a small parameter. The form we shall consider is

$$\epsilon^2\,d^2y/dx^2=f(x)y, \tag{4.82}$$

where ϵ is a small parameter, $f(x)$ is a given function, and y is given at, say, two particular x-values in some range. An important occurrence of (4.82) is the Schrödinger equation in quantum mechanics. We note that, as $\epsilon \to 0$, the equation is singular in the sense that the order decreases from second to zero order when ϵ is set equal to zero. We must therefore expect the solution to be singular as $\epsilon \to 0$. For example, consider

$$\epsilon^2 \, d^2y/dx^2 + y = 0, \tag{4.83}$$

subject to $y(0) = 0$, $y(1) = 1$. This equation has the exact solution

$$y(x) = \frac{\sin(x/\epsilon)}{\sin(1/\epsilon)}. \tag{4.84}$$

Both $\sin(x/\epsilon)$ and $\sin(1/\epsilon)$ rapidly oscillate as $\epsilon \to 0$ for any given x in the range and (4.84) becomes undefined at $\epsilon = 0$.

For equations of the type (4.82), the W.K.B. method gives an approximation which is often close to the exact solution over much of the range. In attempting to find an expression for $y(x)$ as a power series in ϵ, we must have a series which is singular as $\epsilon \to 0$. We adopt the trial expression for $y(x)$ in the form

$$y(x) = \exp\left\{\frac{1}{\epsilon}\int^x [S_0(t) + \epsilon S_1(t) + \epsilon^2 S_2(t) + \ldots]\, dt\right\}, \tag{4.85}$$

which is singular as $\epsilon \to 0$. Here $S_0(t)$, $S_1(t)$, . . . are functions which will be determined by the method. We could have a multiplying constant in front of the exponential in (4.85) if we wished but, instead, by leaving the integral as an indefinite one, we can always fix the lower limit by knowing some initial or boundary conditions on $y(x)$. From (4.85),

$$dy/dx = \frac{1}{\epsilon}[S_0(x) + \epsilon S_1(x) + \epsilon^2 S_2(x) + \ldots]y(x) \tag{4.86}$$

and

$$\frac{d^2y}{dx^2} = \left\{\frac{1}{\epsilon^2}[S_0(x) + \epsilon S_1(x) + \epsilon^2 S_2(x) + \ldots]^2 + \frac{1}{\epsilon}[S_0'(x) + \epsilon S_1'(x) + \epsilon^2 S_2'(x) + \ldots]\right\}y(x). \tag{4.87}$$

Hence, inserting (4.87) into (4.82) and cancelling $y(x)$ on both sides,

$$[S_0(x) + \epsilon S_1(x) + \epsilon^2 S_2(x) + \ldots]^2 + \epsilon[S_0'(x) + \epsilon S_1'(x) + \epsilon^2 S_2'(x) + \ldots] = f(x). \tag{4.88}$$

Equating like powers of ϵ on each side of (4.88) gives an infinite set of equations from which S_0, S_1, S_2, . . . can be determined. The first two of these are

$$S_0^2(x) = f(x), \tag{4.89}$$

$$2S_0(x)S_1(x) + S_0'(x) = 0. \tag{4.90}$$

Hence, from (4.89),

$$S_0(x) = \pm[f(x)]^{\frac{1}{2}}, \tag{4.91}$$

while from (4.90)

$$S_1(x) = -S_0'(x)/2S_0(x). \tag{4.92}$$

In the expression for $y(x)$ in (4.85), we require a term $\exp[\int^x S_1(t)\,dt]$. The integral of S_1 from (4.92) is simply $-\frac{1}{2}\ln[S_0(x)]$ so that

$$\exp\left[\int^x S_1(t)\,dt\right] = [S_0(x)]^{-\frac{1}{2}} = [f(x)]^{-\frac{1}{4}}, \tag{4.93}$$

using (4.91) (apart from constants which again can be combined into the specification of the lower limit of integration).

The W.K.B. approximation consists of neglecting all terms in (4.85) of order ϵ and higher, so that only $S_0(x)$ and $S_1(x)$ are required. Hence, using (4.91) and (4.93), we have the two W.K.B. solutions

$$y(x) = [f(x)]^{-\frac{1}{4}} \exp\left\{\pm\frac{1}{\epsilon}\int^x [f(t)]^{\frac{1}{2}}\,dt\right\}. \tag{4.94}$$

The general solution of (4.82) within the W.K.B. approximation is therefore a linear combination of the two solutions:

$$y(x) = \frac{A}{[f(x)]^{\frac{1}{4}}} \exp\left\{\frac{1}{\epsilon}\int^x [f(t)]^{\frac{1}{2}}\,dt\right\} + \frac{B}{[f(x)]^{\frac{1}{4}}} \exp\left\{-\frac{1}{\epsilon}\int^x [f(t)]^{\frac{1}{2}}\,dt\right\}, \tag{4.95}$$

where A and B are constants.

The solution (4.95) will differ from the exact solution to (4.82) by terms of order ϵ whenever $f(x) \neq 0$. However, if points exist where $f(x) = 0$ (called the turning points of the equation), (4.95) will diverge at these points whereas a numerical integration of (4.82) will give a finite solution there. The W.K.B. method, although a good approximation over much of the range, therefore exhibits the wrong behaviour close to the turning points. We discuss this is more detail in Section 4.8. Solutions which are more accurate away from turning points may be found by including the higher order terms $S_2(x)$, $S_3(x)$ and so on. This will not concern us here.

Example 9 Consider

$$\epsilon^2 \, d^2y/dx^2 = (1+x^2)^2 y, \tag{4.96}$$

subject to $y(0)=0$ and $y(1)=1$. Then $f(x)=(1+x^2)^2$ and (4.95) gives

$$y(x) = \frac{A}{\sqrt{(1+x^2)}} \exp\left[\frac{1}{\epsilon}\int_0^x (1+t^2)\,dt\right] + \frac{B}{\sqrt{(1+x^2)}} \exp\left[-\frac{1}{\epsilon}\int_0^x (1+t^2)\,dt\right], \tag{4.97}$$

where we have chosen the lower limits of integration to be zero. Hence imposing $y(0)=0$ gives

$$0 = A \exp\left[\frac{1}{\epsilon}\,.\,0\right] + B \exp\left[-\frac{1}{\epsilon}\,.\,0\right] \tag{4.98}$$

or

$$A + B = 0. \tag{4.99}$$

Substituting $B=-A$ into (4.97) and performing the integration, we find

$$y(x) = \frac{2A}{\sqrt{(1+x^2)}} \sinh\left[\frac{1}{\epsilon}(x+\tfrac{1}{3}x^3)\right]. \tag{4.100}$$

Applying the boundary condition $y(1)=0$ gives

$$1 = \sqrt{2}\,A \sinh(4/3\epsilon). \tag{4.101}$$

Hence

$$A = 1/\sqrt{2}\sinh(4/3\epsilon) \tag{4.102}$$

and from (4.100)

$$y(x) = \frac{\sqrt{2}}{\sqrt{(1+x^2)}\sinh(4/3\epsilon)} \sinh\left[\frac{1}{\epsilon}(x+\tfrac{1}{3}x^3)\right]. \blacktriangleleft \tag{4.103}$$

Example 10 Consider the Airy-like equation

$$\epsilon^2 \, d^2y/dx^2 + xy = 0 \tag{4.104}$$

(compare (2.209)).

Then $f(x)=-x$ and (4.95) gives

$$y(x) = \frac{A}{x^{\frac{1}{4}}} \exp\left(\frac{\mathrm{i}}{\epsilon}\int^x \sqrt{t}\,dt\right) + \frac{B}{x^{\frac{1}{4}}} \exp\left(-\frac{\mathrm{i}}{\epsilon}\int^x \sqrt{t}\,dt\right), \tag{4.105}$$

whence

$$y(x) = \frac{A}{x^{\frac{1}{4}}} e^{2ix^{3/2}/3\epsilon} + \frac{B}{x^{\frac{1}{4}}} e^{-2ix^{3/2}/3\epsilon} \tag{4.106}$$

$$= Cx^{-\frac{1}{4}} \cos\left(\frac{2}{3\epsilon} x^{\frac{3}{2}} + \delta\right), \tag{4.107}$$

where C and δ are arbitrary constants. ◢

4.7 Eigenvalue problems

An important problem in physics and engineering is that of finding the eigenvalues of a differential equation. Suppose we take

$$d^2y/dx^2 + \lambda y = 0, \tag{4.108}$$

where $y(0) = 0$, $y(\pi) = 0$, and λ is a constant. Then clearly $y = 0$ is a trivial solution. For particular values of λ, non-zero solutions will exist. These values of λ are called the eigenvalues and the corresponding solutions are the eigenfunctions. Now the solution of (4.108) is

$$y(x) = A \cos(x\sqrt{\lambda}) + B \sin(x\sqrt{\lambda}), \tag{4.109}$$

so for $y(0) = 0$ we have $A = 0$, whilst $y(\pi) = 0$ gives

$$B \sin(\pi\sqrt{\lambda}) = 0. \tag{4.110}$$

Now, if we choose $B = 0$, (4.110) is satisfied but, since $A = 0$, the solution for y is the trivial one. Accordingly, we require

$$\sin(\pi\sqrt{\lambda}) = 0 \tag{4.111}$$

for a non-trivial solution. Hence $\sqrt{\lambda} = n$, with $n = 1, 2, 3, \ldots$ and the eigenvalues are therefore

$$\lambda = n^2, \tag{4.112}$$

where $n = 1, 2, 3, \ldots$. The corresponding eigenfunctions are $\sin x$, $\sin(2x)$, $\sin(3x)$, ...

The properties of various special functions discussed and listed in Chapter 2 are often useful in finding the eigenvalues of some types of linear second-order differential equations. For example

$$d^2y/dx^2 + (\lambda - x^2)y = 0, \tag{4.113}$$

where $-\infty < x < \infty$, has non-zero solutions which tend to zero as $x \to \pm\infty$ only when $\lambda = 2n + 1$, $n = 0, 1, 2, \ldots$. These solutions are the Weber–Hermite functions given by $y_n(x) = e^{-x^2/2} H_n(x)$, where $H_n(x)$ are the Hermite polynomials (see (2.200)). There is an infinite number of eigenvalues. Likewise the equation

$$d^2y/dx^2 + (\lambda - x)y = 0, \tag{4.114}$$

where $0 \leqslant x < \infty$, has solutions which are the Airy functions

$$y(x) = Ai(x - \lambda),\ Bi(x - \lambda) \tag{4.115}$$

(see (2.209)–(2.212)). As $x \to \infty$, only the Ai solution tends to zero (the Bi solution tends to infinity as $x \to \infty$). Hence, if we wish to solve (4.114) subject to the boundary conditions

$$y(0) = 0, \quad y(\infty) = 0, \tag{4.116}$$

we see that

$$Ai(-\lambda) = 0. \tag{4.117}$$

This determines the set of λ values which are the eigenvalues of the problem. From tables (see reference on page 22), the λ values are approximately 2.34, 4.09, 5.52, ... There is an infinite number of values of λ for which (4.117) is true (see, for example, the graph of $Ai(x)$ in Figure 2.8).

In general, although we will not prove it here, any differential equation of the form

$$\frac{d}{dx}\left[p(x)\frac{dy}{dx}\right] + [q(x) + \lambda r(x)]y = 0, \tag{4.118}$$

with $y(a) = y(b) = 0$, where a and b are constants (or $\pm\infty$), has an infinite number of real positive eigenvalues provided $p(x) > 0$, $q(x) \leqslant 0$ and $r(x) > 0$. Such a differential equation (with these boundary conditions) is called a Sturm–Liouville system. The calculation of exact eigenvalues is difficult, if not impossible, and approximation techniques are valuable. The following example shows the use of the W.K.B. method in this connection.

Example 11 Consider

$$d^2y/dx^2 + \lambda x^2 y = 0, \tag{4.119}$$

given $y(0) = 0$, $y(\pi) = 0$. Writing this as

$$\frac{1}{\lambda}\frac{d^2y}{dx^2} = -x^2 y \tag{4.120}$$

and comparing with the basic form (4.82), we see that $f(x) = -x^2$ and $\epsilon^2 = 1/\lambda$. Since ϵ is assumed to be small, we expect the method to give the large eigenvalues λ with good accuracy. The general solution of

(4.119) is, from (4.95),

$$y(x)=\frac{A}{\sqrt{x}}\exp\left(i\sqrt{\lambda}\int_0^x t\,dt\right)+\frac{B}{\sqrt{x}}\exp\left(-i\sqrt{\lambda}\int_0^x t\,dt\right) \tag{4.121}$$

or equivalently,

$$y(x)=\frac{C}{\sqrt{x}}\sin(\sqrt{\lambda}\,x^2/2)+\frac{D}{\sqrt{x}}\cos(\sqrt{\lambda}\,x^2/2), \tag{4.122}$$

where A, B, C and D are arbitrary constants. To satisfy the boundary condition $y(0)=0$, we must have $D=0$ in (4.122). The remaining boundary condition $y(\pi)=0$ then gives

$$\frac{C}{\sqrt{\pi}}\sin(\sqrt{\lambda}\,\pi^2/2)=0, \tag{4.123}$$

or

$$\sqrt{\lambda}\,\pi^2/2=n\pi, \tag{4.124}$$

where n is an integer (which must be large in order that λ itself is large). Hence approximate eigenvalues are given by

$$\lambda_n=(2n/\pi)^2, \tag{4.125}$$

for large n. ◢

It should also be mentioned that Sturm–Liouville theory for (4.118) shows that the eigenfunctions y_n can be made, by suitable multiplying constants, to satisfy the relations

$$\int_a^b y_m(x)y_n(x)r(x)\,dx=\delta_{mn}, \tag{4.126}$$

where δ_{mn} is the Kronecker delta (see (1.23)). Such eigenfunctions are said to form an orthonormal set. Since the W.K.B. method of Example 11 gives the eigenfunctions ((4.122) with $D=0$ and λ given by (4.125)) up to an arbitrary multiplicative constant, this constant may be fixed by (4.126). Hence the eigenfunctions and eigenvalues for large n are uniquely determined.

4.8 The Liouville–Green technique

This technique is an attempt to transform the normal form equation

$$d^2y/dx^2=f(x)y \tag{4.127}$$

into another, also in normal form with the same number of turning points, which is soluble exactly or approximately in terms of known functions.

In (4.127), we make a transformation from x to a new variable ξ by writing $x = x(\xi)$, and from $y(x)$ to a new dependent variable $G(\xi)$, where

$$y(x) = \frac{G(\xi)}{(\xi')^{\frac{1}{2}}}, \tag{4.128}$$

with $\xi' = d\xi/dx$. The inclusion of the factor $(\xi')^{-\frac{1}{2}}$ ensures that the equation for $G(\xi)$ is also in normal form, as we now show. From (4.128),

$$\frac{dy}{dx} = (\xi')^{-\frac{1}{2}} \frac{dG}{d\xi} \frac{d\xi}{dx} - \tfrac{1}{2}(\xi')^{-\frac{3}{2}} \xi'' G \tag{4.129}$$

$$= (\xi')^{\frac{1}{2}} \frac{dG}{d\xi} - \tfrac{1}{2}(\xi')^{-\frac{3}{2}} \xi'' G. \tag{4.130}$$

Further,

$$\frac{d^2y}{dx^2} = (\xi')^{\frac{3}{2}} \frac{d^2G}{d\xi^2} + \tfrac{3}{4}(\xi')^{-\frac{5}{2}}(\xi'')^2 G - \tfrac{1}{2}(\xi')^{-\frac{3}{2}} \xi''' G, \tag{4.131}$$

the terms in $dG/d\xi$ cancelling. Inserting (4.131) into (4.127) and using (4.128), we find

$$\frac{d^2G}{d\xi^2} = \left(\frac{f(x)}{\xi'^2} + \Delta\right) G, \tag{4.132}$$

where

$$\Delta = \frac{\xi'''}{2\xi'^3} - \frac{3}{4}\frac{\xi''^2}{\xi'^4} \tag{4.133}$$

is commonly called the Schwarzian derivative. We still have to specify the relationship between x and ξ from which, in principle, the terms in brackets in (4.132) can be expressed in terms of ξ alone. In particular cases this term can be retained in its entirety and (4.132) solved exactly, giving an exact solution to (4.127). In others, the function Δ can be shown to be small compared to the term $f(x)\xi'^{-2}$ and may be neglected, allowing an approximate solution of (4.132), and of (4.127), to be found. We note here that if Δ is neglected and ξ'^2 is chosen to be $f(x)$ in (4.132) then the approximate solutions so generated are precisely those of the W.K.B. approximation. In this case (4.132) becomes

$$d^2G/d\xi^2 = G \tag{4.134}$$

and hence

$$G(\xi) = Ae^{\xi} + Be^{-\xi}. \tag{4.135}$$

Since $\xi'^2 = f(x)$, we have

$$\xi = \int^x [f(t)]^{\frac{1}{2}}\, dt. \tag{4.136}$$

Combining (4.135) and (4.136) with (4.128) gives the W.K.B. result (4.95) (with ϵ formally equal to 1). However, if the equation (4.127) has turning points, that is, if $f(x)$ has zeros in the range of interest, then we clearly should not choose $\xi'^2 = f(x)$ since ξ'^2 is always positive whereas $f(x)$ changes sign. This is the origin of the divergence in the W.K.B. solutions at the turning points of the equation. The correct way to proceed is to choose some function $h(\xi)$ having the same number of zeros as $f(x)$ in the given x-range, and then let

$$h(\xi)\xi'^2 = f(x). \tag{4.137}$$

The choice of $h(\xi)$ should be such that Δ can be neglected in (4.132) and such that the resulting approximate equation for $G(\xi)$,

$$d^2G/d\xi^2 = h(\xi)G, \tag{4.138}$$

is soluble in terms of known functions. We now illustrate various features of this technique.

Example 12 Consider

$$d^2y/dx^2 = e^x y, \tag{4.139}$$

where $-\infty < x < \infty$. Then performing the Liouville–Green transformation (4.128), we find

$$\frac{d^2G}{d\xi^2} = \left(\frac{e^x}{\xi'^2} + \Delta\right)G. \tag{4.140}$$

We now choose

$$\xi'^2 = e^x, \tag{4.141}$$

which gives on integration

$$\xi = 2e^{x/2}. \tag{4.142}$$

Hence, for $-\infty < x < \infty$, we have $0 < \xi < \infty$. From (4.142) the derivatives up to ξ''' can be calculated and substituted into the expression (4.133) for Δ. We find

$$\Delta = -\frac{1}{16e^x} = -\frac{1}{4\xi^2}. \tag{4.143}$$

Hence (4.140) becomes

$$\frac{d^2G}{d\xi^2} = \left(1 - \frac{1}{4\xi^2}\right)G. \tag{4.144}$$

This equation was discussed in Chapter 2 (see (2.157) and (2.159)) where it was shown that the general solution is

$$G(\xi) = A\xi^{\frac{1}{2}}I_0(\xi) + B\xi^{\frac{1}{2}}K_0(\xi), \tag{4.145}$$

where A and B are arbitrary constants and I_0 and K_0 are the modified Bessel functions. Now, from (4.141) and (4.142), $(\xi')^{-\frac{1}{2}} = e^{-x/4} = (2/\xi)^{\frac{1}{2}}$. From the inverse transformation (4.128) relating $G(\xi)$ to $y(x)$, we have finally

$$y(x) = \bar{A}I_0(2e^{x/2}) + \bar{B}K_0(2e^{x/2}), \tag{4.146}$$

where $\bar{A}$ and $\bar{B}$ are arbitrary constants. ◢

Example 13 Consider

$$d^2y/dx^2 = x^4y \tag{4.147}$$

for $x > 0$. Transforming as before, we find

$$\frac{d^2G}{d\xi^2} = \left(\frac{x^4}{\xi'^2} + \Delta\right)G. \tag{4.148}$$

Now choosing

$$\xi'^2 = x^4, \tag{4.149}$$

we have

$$\xi = \tfrac{1}{3}x^3 \tag{4.150}$$

and, from (4.133),

$$\Delta = -2/9\xi^2. \tag{4.151}$$

Hence (4.147) becomes

$$\frac{d^2G}{d\xi^2} = \left(1 - \frac{2}{9\xi^2}\right)G. \tag{4.152}$$

Writing $\xi = \theta/2$, (4.152) becomes

$$\frac{d^2G}{d\theta^2} = \left(\frac{1}{4} - \frac{2}{9\theta^2}\right)G. \tag{4.153}$$

Comparing (4.153) with (2.160) and its solution (2.163), we see that $\nu^2 - \frac{1}{4} = -\frac{2}{9}$ giving $\nu = \frac{1}{6}$, and the solution of (4.153) is therefore

$$G(\theta) = A\sqrt{\theta}\, I_{\frac{1}{6}}(\theta/2) + B\sqrt{\theta}\, K_{\frac{1}{6}}(\theta/2) \tag{4.154}$$

$$= \bar{A}\sqrt{\xi}\, I_{\frac{1}{6}}(\xi) + \bar{B}\sqrt{\xi}\, K_{\frac{1}{6}}(\xi). \tag{4.155}$$

Using the transformation (4.128) to find $y(x)$, we have

$$y(x) = C\sqrt{x}\, I_{\frac{1}{6}}(\tfrac{1}{3}x^3) + D\sqrt{x}\, K_{\frac{1}{6}}(\tfrac{1}{3}x^3). \tag{4.156}$$

We note here that the general form

$$d^2y/dx^2 = x^n y \tag{4.157}$$

for $x > 0$, $n \neq 2$, may be solved by the method of this example. Two simple cases arise:

Case 1 $n = 0$,

$$y = A \sinh x + B \cosh x. \tag{4.158}$$

Case 2 $n = -4$,

$$y = Ax \sinh\left(\frac{1}{x}\right) + Bx \cosh\left(\frac{1}{x}\right). \blacktriangleleft \tag{4.159}$$

Example 14 As we showed in Chapter 3, the Riccati equation

$$dy/dx = y^2 - e^x \tag{4.160}$$

can be transformed to linear form by writing

$$y = -\frac{1}{z}\frac{dz}{dx}. \tag{4.161}$$

After some algebra, (4.160) becomes

$$d^2z/dx^2 = e^x z, \tag{4.162}$$

which can then be solved as in Example 12, and the result substituted into (4.161) to find y. This illustrates the use of the Liouville–Green technique in solving a non-linear equation. ◢

The above examples demonstrated the use of the Liouville–Green technique in solving equations exactly. In many cases this is not possible and we now demonstrate its use in solving equations approximately. We will not discuss equations with no turning points, since, as shown above, the solutions obtained are the same as those of the W.K.B. method. We therefore apply the technique to equations with one or two turning points.

Example 15 Consider

$$d^2y/dx^2 = N^2x(1+x^2)^2y, \tag{4.163}$$

with $-\infty < x < \infty$, and N large. This equation has a single turning point (since the function $x(1+x^2)^2$ changes sign only at $x = 0$). Carrying out the Liouville–Green transformation, we find

$$\frac{d^2G}{d\xi^2} = \left[\frac{N^2x(1+x^2)^2}{\xi'^2} + \Delta\right]G. \tag{4.164}$$

We require (4.164) to have one turning point (if possible) and to be analytically soluble for G if Δ is small enough to be neglected. Choosing

$$\xi'^2\xi = N^2x(1+x^2)^2, \tag{4.165}$$

we obtain the Airy equation

$$d^2G/d\xi^2 = \xi G, \tag{4.166}$$

provided Δ can be neglected. From (4.165), we have

$$\xi' \, |\xi|^{\frac{1}{2}} = N \, |x|^{\frac{1}{2}} \, (1+x^2) \tag{4.167}$$

so that on integration,

$$\tfrac{2}{3} \, |\xi|^{\frac{3}{2}} = N(\tfrac{2}{3} \, |x|^{\frac{3}{2}} + \tfrac{2}{7} \, |x|^{\frac{7}{2}}). \tag{4.168}$$

Hence

$$\xi = N^{\frac{2}{3}}x(1 + \tfrac{3}{7}x^2)^{\frac{2}{3}}. \tag{4.169}$$

The solution of (4.166) is

$$G(\xi) = CAi(\xi) + DBi(\xi), \tag{4.170}$$

and hence the approximate solution for $y(x)$ is found from $y(x) = (\xi')^{-\frac{1}{2}}G(\xi)$ since $(\xi')^{-\frac{1}{2}}$ can be evaluated in terms of x, and ξ can be substituted using (4.169). Further, from (4.169), Δ can be calculated. It is found that Δ is bounded (that is, finite for all x) and that it has an overall factor $N^{-\frac{4}{3}}$, which is a small quantity for large N. It is therefore small over the whole range of x, and may be neglected in (4.164). ◢

For problems in which the equation has two turning points, we consider an important class of eigenvalues problems for which the equation takes the form

$$d^2y/dx^2 = [-E + V(x)]y, \tag{4.171}$$

where $-\infty < x < \infty$, $V(x)$ is finite for finite x, and $y \to 0$ as $x \to \pm\infty$. Here E is the eigenvalue. Transforming equation (4.171), we obtain

$$d^2G/d\xi^2 = \left[\frac{-E + V(x)}{\xi'^2} + \Delta\right]G. \tag{4.172}$$

By assumption, the function $-E + V(x)$ changes sign twice in the range. The simplest equation with two turning points which is soluble in terms of known functions is that satisfied by the Weber–Hermite functions (see (2.206) and (2.207)),

$$d^2G/d\xi^2 = (-\lambda + \xi^2)G, \tag{4.173}$$

where $\lambda = 2n + 1$ (n an integer) for finite G as $\xi \to \pm\infty$. Hence we choose

$$\xi'^2(-\lambda + \xi^2) = -E + V(x) \tag{4.174}$$

in (4.172). Neglecting Δ, (4.172) becomes (4.173) which can be solved in terms of the Weber–Hermite functions. We note, however, that on taking the square root and integrating (4.174), the relationship between ξ and x can only be found implicitly (that is, we cannot write ξ explicitly as a function of x). Nevertheless, an approximate formula giving the eigenvalues E_n can be obtained from (4.174). The turning points, say x_1 and x_2, correspond to $\xi = \pm\sqrt{\lambda}$. Hence integrating (4.174) between the turning points after taking the square root gives

$$\int_{-\sqrt{\lambda}}^{+\sqrt{\lambda}} \sqrt{(\xi^2 - \lambda)}\, d\xi = \int_{x_1}^{x_2} \sqrt{[E - V(x)]}\, dx. \tag{4.175}$$

Performing the integration on the left-hand side and substituting $\lambda = 2n + 1$, we find

$$\int_{x_1}^{x_2} \sqrt{[E_n - V(x)]}\, dx = \pi(n + \tfrac{1}{2}). \tag{4.176}$$

This is known as the Bohr–Sommerfeld formula. It can also be derived within the W.K.B. approximation using so-called connection formulae which are not discussed here. For each value of n, a value of E_n can be found (if necessary, numerically) and this value is an approximation to the corresponding exact eigenvalue of the equation. In fact, for a particular class of potentials $V(x)$, it can be shown that the formula provides the exact eigenvalues of the equation.

Problems 4

1. Find the first three non-zero terms in the series solution of the Airy equation

$$d^2y/dx^2 = xy,$$

given $y = 0$ and $dy/dx = 1$ at $x = 0$.

2. Find the Frobenius series solutions of the differential equation

$$3x\frac{d^2y}{dx^2}-(1-x)\frac{dy}{dx}-y=0$$

in the form

$$y=\sum_{r=0}^{\infty} a_r x^{m+r}.$$

Show that, if $y(0)=0$, then

$$y=Ax^{\frac{4}{3}}\left(1-\frac{x}{21}+\frac{x^2}{315}-\ldots\right)$$

is the solution (A arbitrary).

3. Obtain by Picard's method the first three terms of the series solution of the equation

$$dy/dx=1+2xy+y^2,$$

given $y(0)=0$.

4. Obtain by Picard's method the first three approximations to the solution of the simultaneous equations

$$dy/dx=2x+z,$$
$$dz/dx=3xy+x^2z,$$

given $y(0)=2$, $z(0)=0$.

5. Use the perturbation series method of Example 7 to obtain a solution, to first order in the small parameter μ, of the van der Pol equation

$$\frac{d^2y}{dt^2}+\mu(y^2-1)\frac{dy}{dt}+y=0,$$

taking $y_0(t)=\cos t$.

6. Reduce the equation

$$\frac{d^2y}{dx^2}+x^2\frac{dy}{dx}+(4+x+\tfrac{1}{4}x^4)y=0$$

to normal form, and hence find its general solution.

7. Use the W.K.B. method to find approximate solutions of the equation

$$d^2y/dx^2=N^2x^4y.$$

8. Use the Liouville–Green transformation to obtain solutions of

$$d^2y/dx^2=x^ny$$

for $x>0$, where n is a positive constant, in terms of the modified Bessel functions. (See Example 13.)

5
Contour integration

5.1 Functions of a complex variable

It will be necessary in this chapter to obtain a number of important results, the applications of which will be discussed in Chapter 6. We begin with some elementary ideas and lead up to the main topic of complex integration in Section 5.6.

A complex number has the form

$$z = x + \mathrm{i}y, \tag{5.1}$$

where x and y are real and $\mathrm{i}^2 = -1$, and may be represented geometrically in the complex (Argand) plane as shown in Figure 5.1. We shall subsequently consider functions of the complex number z. One of the many reasons for wanting to do this may be illustrated by the following example.

Consider the function $f(x) = (1 + x^2)^{-1}$, where x is real. The power series expansion of this function in powers of x only converges for $|x| < 1$. This may seem strange since $f(x)$ is finite and well-behaved for all values of real x. However, in the complex plane, $f(z) = (1 + z^2)^{-1}$ diverges as z approaches the points $+\mathrm{i}$ and $-\mathrm{i}$ which are at unit distance from the origin. For all z inside the circle $|z| = 1, f(z)$ is well behaved. The radius of convergence of the power series expansion in z of $f(z)$ is therefore unity. Hence, when $z = x$ (real), the radius of convergence is also unity. The behaviour of $f(x)$ is more clearly understood by examining that of $f(z)$ in the complex plane.

From Figure 5.1, an alternative form for z is obtained by writing

$$x = r \cos \theta, \quad y = r \sin \theta, \tag{5.2}$$

where $0 \leqslant \theta < 2\pi$ or any other interval of width 2π which includes $\theta = 0$, for example $-\pi < \theta \leqslant \pi$. Using Euler's formula, we have

$$z = r(\cos\theta + \mathrm{i}\sin\theta) = r\mathrm{e}^{\mathrm{i}(\theta+2n\pi)}, \tag{5.3}$$

where $n = 0, \pm 1, \pm 2, \ldots,$ since each time we change θ by 2π we return to the original point. In (5.1), x and y are called the real and imaginary parts of z and are denoted $\mathrm{Re}\, z$ and $\mathrm{Im}\, z$, respectively; $r = |z| = \sqrt{(x^2 + y^2)}$ is the modulus (or magnitude) of z and θ is the phase or argument of z, denoted by $\arg z$. We note here that the set of points x, y satisfying, say, $|z| = 4$ form a circle, centre the origin, radius 4, whilst those points x, y satisfying, say, $|z - 2| = 1$ form a circle, centre $(2, 0)$, radius 1.

The complex conjugate of z is denoted by z^* (or sometimes by $\bar{z}$) and is defined by changing the sign of i in (5.1) to give

$$z^* = x - \mathrm{i}y. \tag{5.4}$$

Consequently, using (5.1) and (5.4),

$$x = \tfrac{1}{2}(z + z^*), \quad y = \frac{1}{2\mathrm{i}}(z - z^*). \tag{5.5}$$

We define a function of z, $f(z)$, to have the form

$$\omega = f(z) = u(x, y) + \mathrm{i}v(x, y), \tag{5.6}$$

where $u(x, y)$, $v(x, y)$ are real functions. In general we have the modulus of ω given by

$$|\omega| = |f(z)| = \sqrt{(u^2 + v^2)}. \tag{5.7}$$

Furthermore, ω is single-valued if to each value of z (in some region of the complex plane) there is precisely one value of ω. Consider the following standard examples:

$$\omega = z^2 = (x + \mathrm{i}y)^2 = (x^2 - y^2) + \mathrm{i}(2xy) \tag{5.8}$$

Figure 5.1

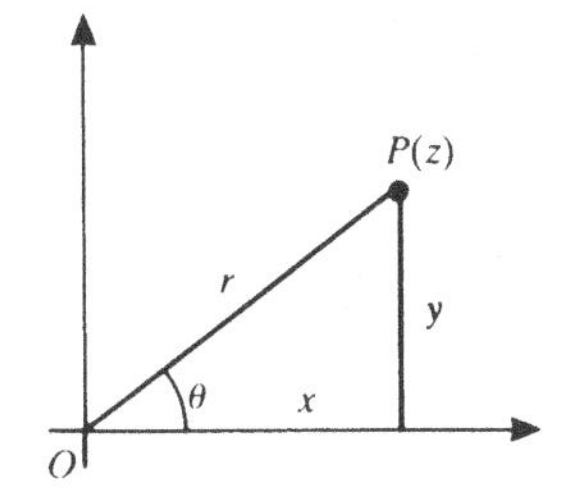

or equivalently

$$\omega = z^2 = r^2 e^{2i(\theta+2n\pi)}, \tag{5.9}$$

is a single-valued function of z. On the other hand, $\omega = \sqrt{z}$ $(z \neq 0)$ is not a single-valued function since, writing $z = re^{i(\theta+2n\pi)}$, we have

$$\omega = \sqrt{z} = \sqrt{r}\, e^{i(\theta+2n\pi)/2}. \tag{5.10}$$

The values $n = 0$ and 1 give two distinct solutions

$$\omega_1 = \sqrt{r}\, e^{i\theta/2} = \sqrt{r}\,[\cos(\theta/2) + i\sin(\theta/2)] \tag{5.11}$$

and

$$\omega_2 = \sqrt{r}\, e^{i(\theta/2+\pi)} = \sqrt{r}\,[\cos(\theta/2 + \pi) + i\sin(\theta/2 + \pi)]. \tag{5.12}$$

These two functions are called the two branches of $\omega = \sqrt{z}$, and $\omega = \sqrt{z}$ is termed a multi-valued function. If we give r and θ a succession of different values then ω_1 traces out some curve in the complex plane; ω_2 is obtained by rotating the curve through π about the origin according to (5.11) and (5.12) (see Figure 5.2). Two values of $\sqrt{z}$ exist for all z except $z = 0$, which is called the branch point. It is important to realize that the two curves in Figure 5.2 are not the graphs of $\omega = \sqrt{z}$ but only the paths along which z moves as we change r and θ (or equivalently x and y). By analogy with the above, $\omega = z^{\frac{1}{3}}$ will have three branches with a branch point at $z = 0$, while $\omega = \sqrt{[(z - a)(z - b)]}$ will have two branches with two branch points at $z = a$ and $z = b$. We shall expand the discussion of branch points and multi-valued functions in Chapter 6.

Figure 5.2

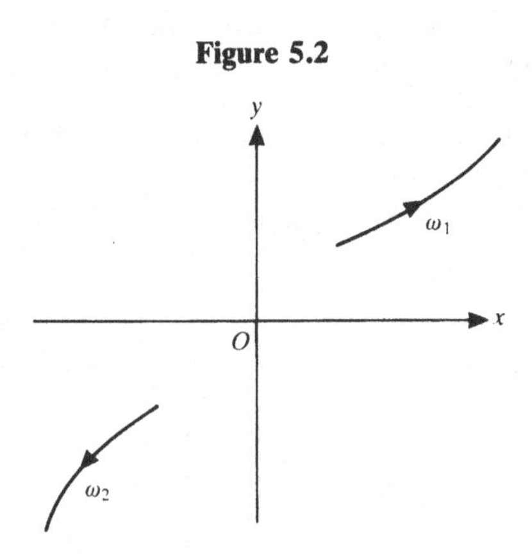

5.2 Exponential and logarithmic functions

We define

$$e^z = e^{(x+iy)} = e^x e^{iy} = e^x(\cos y + i \sin y), \tag{5.13}$$

using Euler's formula. This definition satisfies the obvious requirement that e^z should become e^x when z is real. Further

$$e^z = 1 + z + \frac{z^2}{2!} + \frac{z^3}{3!} + \dots. \tag{5.14}$$

From (5.13) when $x = 0$,

$$e^{iy} = \cos y + i \sin y \tag{5.15}$$

and consequently

$$e^{-iy} = \cos y - i \sin y. \tag{5.16}$$

It follows from (5.15) and (5.16) that

$$\cos y = \tfrac{1}{2}(e^{iy} + e^{-iy}), \tag{5.17}$$

$$\sin y = \frac{1}{2i}(e^{iy} - e^{-iy}), \tag{5.18}$$

where y is a real variable. We may now extend these formulae to define the trigonometric functions of a complex variable z by

$$\cos z = \tfrac{1}{2}(e^{iz} + e^{-iz}), \tag{5.19}$$

$$\sin z = \frac{1}{2i}(e^{iz} - e^{-iz}). \tag{5.20}$$

From (5.19) and (5.20), we can immediately define $\tan z$, $\sec z$, $\operatorname{cosec} z$, $\cot z$ as in the case of real variable theory. Similarly, the hyperbolic functions $\cosh z$ and $\sinh z$ can be defined as

$$\cosh z = \tfrac{1}{2}(e^z + e^{-z}), \tag{5.21}$$

$$\sinh z = \tfrac{1}{2}(e^z - e^{-z}), \tag{5.22}$$

which reduce to the usual definitions when z is real. The other hyperbolic functions are defined in an obvious way. It is readily seen that

$$\sin^2 z + \cos^2 z = 1, \tag{5.23}$$

$$\cosh^2 z - \sinh^2 z = 1, \tag{5.24}$$

with the other identities of real variable theory remaining true when the real variable x is replaced by the complex variable z.

Closely related to the exponential function e^z is the logarithmic function $\ln z$. If we write

$$z = e^{\omega} \tag{5.25}$$

then

$$\omega = \ln z = \ln(re^{i(\theta+2n\pi)}), \tag{5.26}$$

from (5.3). Hence

$$\omega = \ln r + i(\theta + 2n\pi), \tag{5.27}$$

where $n = 0, \pm 1, \pm 2, \ldots$, and we see that $\omega = \ln z$ is a multivalued function. Writing $\omega = u + iv$, we have

$$u = \ln r, \quad v = \theta + 2n\pi \tag{5.28}$$

for the real and imaginary parts of ω. When $n = 0$ and $-\pi < \theta \leqslant \pi$, ω takes its principal value and $\omega = \ln z$ is then a single-valued function. Since $r = |z|$ and $\theta = \arg z$, we can write

$$\omega = \ln |z| + i(\arg z + 2n\pi). \tag{5.29}$$

Sometimes, when the principal value of ω is taken, we write $\omega = \mathrm{Ln}\, z$, the capital letter L meaning that n has been put equal to zero in (5.29). We now give examples and include some elementary results.

Example 1

(*a*)
$$\ln(-2) = \ln |-2| + i(\arg(-2) + 2n\pi). \tag{5.30}$$

Hence, since $-2 = 2e^{\pi i}$, $|-2| = 2$ and $\arg(-2) = \pi$, and therefore

$$\ln(-2) = \ln 2 + i(\pi + 2n\pi). \tag{5.31}$$

(*b*)
$$\mathrm{Ln}(-i) = \ln |-i| + i \arg(-i). \tag{5.32}$$

Hence, since $-i = 1e^{-\pi i/2}$, $|-i| = 1$ and $\arg(-i) = -\pi/2$, and therefore

$$\mathrm{Ln}(-i) = \ln 1 + i(-\pi/2) = -\pi i/2. \tag{5.33}$$

(Similarly $\mathrm{Ln}(i) = \pi i/2$).
(c) Complex powers α of z can be established using the relation

$$z^{\alpha} = e^{\alpha \ln z}, \tag{5.34}$$

where $\ln z$ is defined by (5.27). For example the principal value of i^i is given by

$$i^i = e^{i\,\mathrm{Ln}\, i} = e^{i(\pi i/2)} = e^{-\pi/2}, \tag{5.35}$$

which (surprisingly perhaps) is *real.* ◢

Example 2 To determine all possible values of

$$\omega = \sin^{-1} z \tag{5.36}$$

when $z = 2$. (We note that for real variables $\sin^{-1} 2$ makes no sense since the sine function is bounded by ± 1.) Now

$$z = \sin \omega = \frac{e^{i\omega} - e^{-i\omega}}{2i} = \frac{e^{2i\omega} - 1}{2ie^{i\omega}}. \tag{5.37}$$

Hence

$$e^{2i\omega} - 2ie^{i\omega}z - 1 = 0. \tag{5.38}$$

This is a quadratic equation for the quantity $e^{i\omega}$, so that

$$e^{i\omega} = \tfrac{1}{2}[2iz \pm \sqrt{(-4z^2 + 4)}] = iz \pm \sqrt{(1 - z^2)}. \tag{5.39}$$

Taking the logarithm of (5.39), we find

$$i\omega = \ln[iz \pm \sqrt{(1 - z^2)}]. \tag{5.40}$$

Now, putting $z = 2$,

$$i\omega = \ln[(2 \pm \sqrt{3})i]. \tag{5.41}$$

Hence

$$\omega = \frac{1}{i}\left[\ln(2 \pm \sqrt{3}) + i\left(\frac{\pi}{2} + 2n\pi\right)\right], \tag{5.42}$$

since the points $(2 \pm \sqrt{3})i$ have arguments $\pi/2$. Finally then

$$\omega = \frac{\pi}{2} + 2n\pi - i\ln(2 \pm \sqrt{3}). \tag{5.43}$$

Since $(2 - \sqrt{3})(2 + \sqrt{3}) = 1$,

$$\ln(2 - \sqrt{3}) + \ln(2 + \sqrt{3}) = 0 \tag{5.44}$$

and therefore (5.43) may be written

$$\omega = \frac{\pi}{2} + 2n\pi \pm i\ln(2 + \sqrt{3}). \quad ◢ \tag{5.45}$$

5.3 The derivative of a complex function

We say that a function $f(z)$ is *analytic* at a point z_0 if it has a unique derivative at that point. By comparison with the real variable definition of a derivative, we now write

$$f'(z_0) = \left.\frac{df}{dz}\right|_{z=z_0} = \lim_{\delta z \to 0} \frac{f(z_0 + \delta z) - f(z_0)}{\delta z}. \tag{5.46}$$

Equivalently, we have

$$f'(z_0) = \lim_{z \to z_0} \frac{f(z) - f(z_0)}{z - z_0}. \tag{5.47}$$

If in addition $f(z_0)$ has a single finite value, and $f'(z_0)$ is unique, then the function is said to be *regular* at $z = z_0$. In (5.46) and (5.47), the limit must be independent of the path along which $\delta z \to 0$ (that is, $z \to z_0$). Consider Figure 5.3 and assume $f(z)$ is defined in some (shaded) region R. Then if $z = z_0 + \delta z$ is some point in this region it can be made to approach z_0 along any one of the infinite number of paths joining z to z_0 (C_1, C_2, and C_3 are three such paths). If $f'(z_0)$ is to have a unique value, then (5.46) and (5.47) must be independent of the way z approaches z_0 (or $\delta z \to 0$). We note that in real variable theory we do not have the freedom that complex variable theory produces for, if x_0 is some point on the x-axis, a neighbouring point x may approach x_0 in only two possible ways – from the left and from the right. The requirement that the derivative be independent of the path in the Argand plane restricts the type of functions which possess a derivative. On the assumption that $f(z)$ does have a unique derivative at z_0, it follows from (5.47) that it must be continuous at z_0 in the sense that

$$f(z_0 + \delta z) \to f(z_0) \quad \text{as} \quad \delta z \to 0 \tag{5.48}$$

along any path. Functions which do not satisfy (5.48) are discontinuous at z_0; z_0 is then said to be a singularity of the function. We shall discuss this concept in more detail later.

Figure 5.3

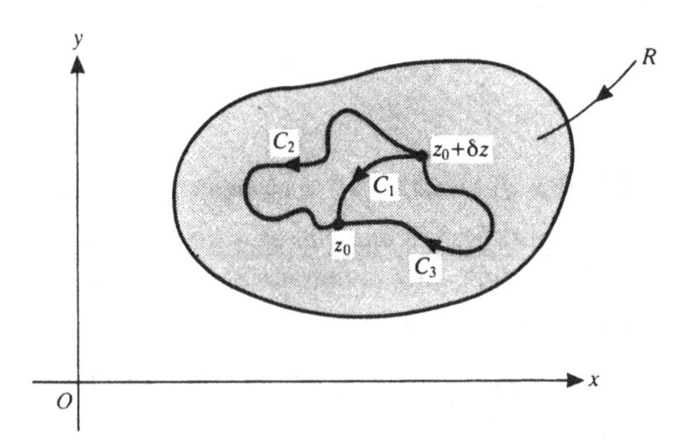

5.4 The Cauchy–Riemann equations

We now obtain the conditions under which a function

$$\omega = f(z) = u(x, y) + \mathrm{i}v(x, y) \tag{5.49}$$

has a unique derivative as given by (5.46) or (5.47). Writing $\delta z = \delta x + \mathrm{i}\delta y$ and $z_0 = x_0 + \mathrm{i}y_0$ then

$$\begin{aligned} f(z_0 + \delta z) - f(z_0) &= [u(x_0 + \delta x, y_0 + \delta y) - u(x_0, y_0)] \\ &\quad + \mathrm{i}[v(x_0 + \delta x, y_0 + \delta y) - v(x_0, y_0)] \qquad (5.50) \\ &= \delta u + \mathrm{i}\delta v. \qquad (5.51) \end{aligned}$$

Hence

$$f'(z_0) = \lim_{\delta z \to 0} \frac{\delta u + \mathrm{i}\delta v}{\delta z} = \lim_{\substack{\delta x \to 0 \\ \delta y \to 0}} \left(\frac{\delta u + \mathrm{i}\delta v}{\delta x + \mathrm{i}\delta y} \right). \tag{5.52}$$

If the limit is to be path-independent (as required), we can take the limit in two particular ways: (i) allowing $\delta x \to 0$ and then $\delta y \to 0$ (path C_1 in Figure 5.4), and (ii) allowing $\delta y \to 0$ followed by $\delta x \to 0$ (path C_2 in Figure 5.4). We require that the two results should be identical. As $\delta y \to 0$, we have from (5.52)

$$f'(z_0) = \lim_{\delta x \to 0} \left(\frac{\delta u + \mathrm{i}\delta v}{\delta x} \right) = \frac{\partial u}{\partial x} + \mathrm{i}\frac{\partial v}{\partial x}, \tag{5.53}$$

whereas when $\delta x \to 0$ we have

$$f'(z_0) = \lim_{\delta y \to 0} \left(\frac{\delta u + \mathrm{i}\delta v}{\mathrm{i}\delta y} \right) = \frac{\partial v}{\partial y} - \mathrm{i}\frac{\partial u}{\partial y}. \tag{5.54}$$

For (5.53) and (5.54) to be identical, we require

$$\frac{\partial u}{\partial x} + \mathrm{i}\frac{\partial v}{\partial x} = \frac{\partial v}{\partial y} - \mathrm{i}\frac{\partial u}{\partial y}, \tag{5.55}$$

Figure 5.4

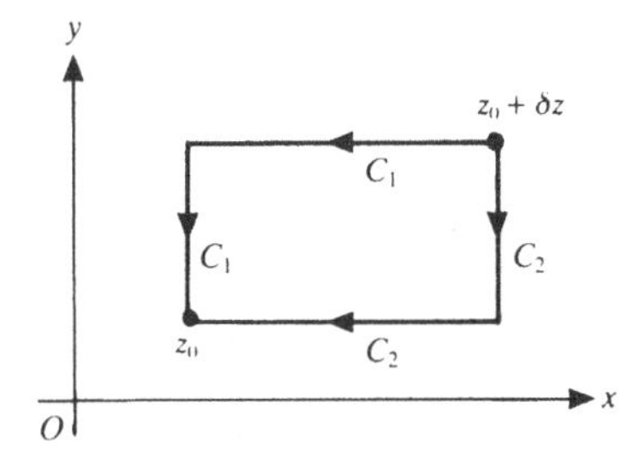

or equating real and imaginary parts (remembering that u and v are real functions), we find

$$\partial u/\partial x = \partial v/\partial y, \quad \partial u/\partial y = -\partial v/\partial x. \tag{5.56}$$

The equations (5.56) are called the Cauchy–Riemann equations and are the necessary conditions for the limit expression for the derivative to have a unique value. However, these conditions are not *sufficient* since we have considered only two possible paths out of the infinity of possible paths joining $z_0 + \delta z$ to z_0. It can be proved (although we shall not do this) that provided the first-order partial derivatives $\partial u/\partial x$, $\partial v/\partial x$, $\partial u/\partial y$, $\partial v/\partial y$ are continuous at z_0 then (5.56) are necessary *and* sufficient for the limit expression to be path-independent and for the derivative to be unique. We now give some examples of the use of the Cauchy–Riemann equations.

Example 3 The function

$$\omega = z^2 = (x + \mathrm{i}y)^2 = (x^2 - y^2) + 2\mathrm{i}xy \tag{5.57}$$

is analytic since $u = x^2 - y^2$, $v = 2xy$, and

$$\partial u/\partial x = 2x = \partial v/\partial y, \tag{5.58}$$

$$\partial u/\partial y = -2y = -\partial v/\partial x. \tag{5.59}$$

The Cauchy–Riemann equations are satisfied and the first derivatives are continuous. Hence

$$\frac{d\omega}{dz} = \frac{\partial u}{\partial x} + \mathrm{i}\frac{\partial v}{\partial x} = 2x + \mathrm{i}2y = 2z. \tag{5.60}$$

This is precisely the result which would have been obtained by differentiating $\omega = z^2$ assuming ω and z were real. In fact, the usual rules of differentiation are still valid when we are dealing with complex analytic functions. For example, if $G = F(\omega)$ and $\omega = f(z)$ then

$$\frac{dG}{dz} = \frac{dF}{d\omega}\frac{d\omega}{dz}. \quad ◢ \tag{5.61}$$

Example 4 Consider

$$\omega = \mathrm{e}^z = \mathrm{e}^{x+\mathrm{i}y} = \mathrm{e}^x(\cos y + \mathrm{i}\sin y). \tag{5.62}$$

Then

$$u = \mathrm{e}^x \cos y, \quad v = \mathrm{e}^x \sin y \tag{5.63}$$

and

$$\partial u/\partial x = \mathrm{e}^x \cos y = \partial v/\partial y, \tag{5.64}$$

$$\partial u/\partial y = -\mathrm{e}^x \sin y = -\partial v/\partial x. \tag{5.65}$$

Again the Cauchy–Riemann equations are satisfied and the derivatives in (5.64) and (5.65) are continuous. Hence $\omega = \mathrm{e}^z$ is an analytic function and

$$\frac{d\omega}{dz} = \frac{\partial u}{\partial x} + \mathrm{i}\frac{\partial v}{\partial x} = \mathrm{e}^x \cos y + \mathrm{i}\mathrm{e}^x \sin y \tag{5.66}$$

$$= \mathrm{e}^{x+\mathrm{i}y} = \mathrm{e}^z. \tag{5.67}$$

Consequently, as expected,

$$\frac{d}{dz}(\mathrm{e}^z) = \mathrm{e}^z. \tag{5.68}$$

Example 5 Consider

$$\omega = 1/z. \tag{5.69}$$

Then

$$\omega = \frac{1}{x + \mathrm{i}y} = \frac{x - \mathrm{i}y}{x^2 + y^2} = \frac{x}{x^2 + y^2} - \mathrm{i}\frac{y}{x^2 + y^2}. \tag{5.70}$$

Hence

$$u = \frac{x}{x^2 + y^2}, \quad v = \frac{-y}{x^2 + y^2} \tag{5.71}$$

and

$$\frac{\partial u}{\partial x} = \frac{y^2 - x^2}{(x^2 + y^2)^2} = \frac{\partial v}{\partial y}, \tag{5.72}$$

$$\frac{\partial u}{\partial y} = \frac{-2xy}{(x^2 + y^2)^2} = -\frac{\partial v}{\partial x}. \tag{5.73}$$

Therefore, the Cauchy–Riemann equations are satisfied except when $x = y = 0$ (since the derivatives in (5.72) and (5.73) are not defined there). Consequently, $\omega = 1/z$ is an analytic function everywhere except at $z = 0$, this point being termed a singularity of the function.

Example 6 The last example illustrated a function which is analytic everywhere except at one point. Some functions are not analytic anywhere. Consider

$$\omega = z^* = x - \mathrm{i}y. \tag{5.74}$$

Then $u = x$ and $v = -y$, so

$$\partial u/\partial x = 1, \quad \partial v/\partial y = -1, \quad \partial u/\partial y = 0, \quad \partial v/\partial x = 0. \tag{5.75}$$

Consequently, $\partial u/\partial x \neq \partial v/\partial y$ and the function is not analytic for any x and y.

Similarly, consider

$$\omega = |z|^2 = zz^* = (x + \mathrm{i}y)(x - \mathrm{i}y) = x^2 + y^2. \qquad (5.76)$$

Hence $u = x^2 + y^2$ and $v = 0$ and the Cauchy–Riemann equations are therefore not satisfied except at $x = y = 0$. These results clearly imply that the limit in the expression for the derivative depends on the path along which $z \to z_0$ (or $\delta z \to 0$). To find the derivative of $\omega = |z|^2$ (see (5.47)) we need the limit of

$$\frac{|z|^2 - |z_0|^2}{z - z_0} = \frac{zz^* - z_0 z_0^*}{z - z_0} \qquad (5.77)$$

$$= z^* + \frac{z_0 z^* - z_0 z_0^*}{z - z_0} \qquad (5.78)$$

$$= z^* + z_0\left(\frac{z^* - z_0^*}{z - z_0}\right). \qquad (5.79)$$

Hence, if $z - z_0 = r\mathrm{e}^{\mathrm{i}\theta}$,

$$\frac{z^* - z_0^*}{z - z_0} = \frac{(z - z_0)^*}{z - z_0} = \frac{r\mathrm{e}^{-\mathrm{i}\theta}}{r\mathrm{e}^{\mathrm{i}\theta}} = \mathrm{e}^{-2\mathrm{i}\theta}, \qquad (5.80)$$

and

$$\frac{|z|^2 - |z_0|^2}{z - z_0} = z^* + z_0[\cos(2\theta) - \mathrm{i}\sin(2\theta)]. \qquad (5.81)$$

As $z \to z_0$, the limit depends on θ, the angle at which z approaches z_0. ◢

5.5 Derivatives of multi-valued functions

When a function is not single-valued (for example $\sqrt{z}$ and $\ln z$) then the derivative may be obtained only on a particular branch (see Section 5.2). If, for example, $\omega = \ln z = \ln r + \mathrm{i}(\theta + 2n\pi)$, then by taking a particular value of n (say $n = 0$) we have a single-valued function $z = \mathrm{e}^{\omega}$. For this branch

$$dz/d\omega = \mathrm{e}^{\omega} \qquad (5.82)$$

which implies

$$d\omega/dz = \mathrm{e}^{-\omega} = 1/z, \qquad (5.83)$$

except at $z = 0$ ($\ln z$ is not analytic at $z = 0$).

5.6 Complex integration

We first indicate the meaning to be attached to $\int_C f(z)\,dz$, where C is some path in the complex plane with end points $A(z_1)$ and $B(z_2)$ (see Figure 5.5). Since $f(z)$ may be written as $f(z) = u(x, y) + \mathrm{i}v(x, y)$ and $dz = dx + \mathrm{i}\,dy$, then $f(z)\,dz = (u + \mathrm{i}v)(dx + \mathrm{i}\,dy)$ and

$$\int_C f(z)\,dz = \int_C (u\,dx - v\,dy) + \mathrm{i}\int_C (v\,dx + u\,dy). \qquad (5.84)$$

Each integral on the right of (5.84) is a real line integral and may be evaluated by integrating from A to B along the specified path C (the equation of which will, in general, be given as $y = y(x)$, say). Since the complex integral can be put in the form (5.84), the usual rules of integration which apply to real integrals must also apply to complex ones. Accordingly, we have

(i)
$$\int_C [f(z) + g(z)]\,dz = \int_C f(z)\,dz + \int_C g(z)\,dz, \qquad (5.85)$$

where f and g are any two integrable functions;

(ii)
$$\int_C Kf(z)\,dz = K\int_C f(z)\,dz, \qquad (5.86)$$

where K is an arbitrary complex constant;

(iii)
$$\int_{AEB} f(z)\,dz = \int_{AE} f(z)\,dz + \int_{EB} f(z)\,dz, \qquad (5.87)$$

where E is some point on the curve C lying between A and B (see Figure 5.5).

An important result follows if $f(z)$ is analytic in a region R

Figure 5.5

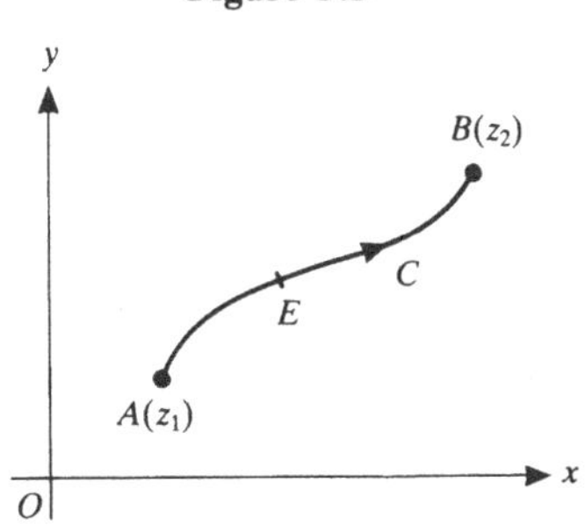

containing the points $A(z_1)$ and $B(z_2)$. In this case we can show that

$$\int_{z_1}^{z_2} f'(z)\,dz = [f(z)]_{z_1}^{z_2} = f(z_2) - f(z_1), \tag{5.88}$$

which implies that the integral is independent of the path C since the result depends only on the values of $f(z)$ at the two end points. Consider

$$\int_{z_1}^{z_2} f'(z)\,dz = \int_{z_1}^{z_2} \left(\frac{\partial u}{\partial x} + \mathrm{i}\frac{\partial v}{\partial x}\right)(dx + \mathrm{i}\,dy), \tag{5.89}$$

using the form for $f'(z)$ given in (5.53). Then

$$\int_{z_1}^{z_2} f'(z)\,dz = \int_{z_1}^{z_2} \left(\frac{\partial u}{\partial x}dx - \frac{\partial v}{\partial x}dy\right) + \mathrm{i}\int_{z_1}^{z_2} \left(\frac{\partial v}{\partial x}dx + \frac{\partial u}{\partial x}dy\right). \tag{5.90}$$

Now, since $f(z)$ is assumed to be analytic, the Cauchy–Riemann equations

$$\partial u/\partial x = \partial v/\partial y, \quad \partial u/\partial y = -\partial v/\partial x \tag{5.91}$$

hold. Consequently the right-hand side of (5.90) becomes

$$\int_{z_1}^{z_2} \left(\frac{\partial u}{\partial x}dx + \frac{\partial u}{\partial y}dy\right) + \mathrm{i}\int_{z_1}^{z_2} \left(\frac{\partial v}{\partial x}dx + \frac{\partial v}{\partial y}dy\right) \tag{5.92}$$

$$= \int_{z_1}^{z_2} du + \mathrm{i}\int_{z_1}^{z_2} dv, \tag{5.93}$$

since the integrands in (5.92) are the total derivatives of u and v. Finally, (5.93) integrates to give

$$\int_C f'(z)\,dz = [u]_{z_1}^{z_2} + \mathrm{i}[v]_{z_1}^{z_2} = [u + \mathrm{i}v]_{z_1}^{z_2}$$

$$= [f(z)]_{z_1}^{z_2} = f(z_2) - f(z_1). \tag{5.94}$$

This proves the result that the integral is independent of the path. The following examples illustrate these ideas.

Example 7 Evaluate

$$\int_C z^2\,dz, \tag{5.95}$$

where

(i) $C = C_1$ is the straight line $y = x$ from $A(0, 0)$ to $B(1, 1)$,
(ii) $C = C_2$ is the x-axis from $(0, 0)$ to $A(1, 0)$ and the straight line from $A(1, 0)$ to $B(1, 1)$,

(see Figure 5.6).

(i)
$$\int_{C_1} z^2\,dz = \int_{C_1} (x^2 - y^2 + 2ixy)(dx + i dy)$$
$$= \int_{C_1} [(x^2 - y^2)\,dx - 2xy\,dy]$$
$$+ i \int_{C_1} [2xy\,dx + (x^2 - y^2)\,dy]. \tag{5.96}$$

Now on C_1, $y = x$. Inserting this into (5.96) we find (using $dy = dx$)

$$\int_{C_1} z^2\,dz = \int_0^1 [(x^2 - x^2)\,dx - 2x^2\,dx]$$
$$+ i \int_0^1 [2x^2\,dx + (x^2 - x^2)\,dx] \tag{5.97}$$
$$= (i - 1) \int_0^1 2x^2\,dx = \tfrac{2}{3}(-1 + i). \tag{5.98}$$

(ii)
$$\int_{C_2} z^2\,dz = \int_{C_2} [(x^2 - y^2)\,dx - 2xy\,dy]$$
$$+ i \int_{C_2} [2xy\,dx + (x^2 - y^2)\,dy], \tag{5.99}$$

as in (5.96). Splitting this into two parts, one for each section of the path C_2, we have

$$\int_{C_2} z^2\,dz = \int_{(0,0)}^{(1,0)} [(x^2 - y^2)\,dx - 2xy\,dy]$$
$$+ i \int_{(0,0)}^{(1,0)} [2xy\,dx + (x^2 - y^2)\,dy]$$
$$+ \int_{(1,0)}^{(1,1)} [(x^2 - y^2)\,dx - 2xy\,dy]$$
$$+ i \int_{(1,0)}^{(1,1)} [2xy\,dx + (x^2 - y^2)\,dy], \tag{5.100}$$

Figure 5.6

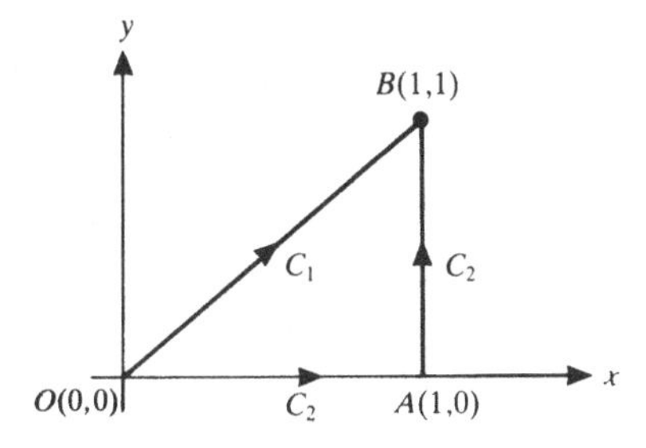

where the integrals are along the straight lines joining the end points. Along OA, $y = 0$ and hence $dy = 0$. The first two integrals in (5.100) therefore give

$$\int_0^1 x^2\,dx + \mathrm{i}\int_0^1 0\,dx = \tfrac{1}{3}. \tag{5.101}$$

Along AB, $x = 1$ and $dx = 0$ and the second pair of integrals in (5.100) gives

$$\int_0^1 -2y\,dy + \mathrm{i}\int_0^1 (1 - y^2)\,dy = -1 + \tfrac{2}{3}\mathrm{i}. \tag{5.102}$$

Hence

$$\int_{C_2} z^2\,dz = \tfrac{1}{3} - 1 + \tfrac{2}{3}\mathrm{i} = \tfrac{2}{3}(-1 + \mathrm{i}). \tag{5.103}$$

The results (5.98) and (5.103) are the same. This is to be expected since $z^3/3$ (the function which differentiates to give z^2) is an analytic function and hence the value of the integral is $\frac{1}{3}(1 + \mathrm{i})^3 = \frac{2}{3}(-1 + \mathrm{i})$. ◢

5.7 Cauchy's Theorem (First Integral Theorem)

This may be stated as follows: If $f(z)$ is single-valued and analytic throughout a simply connected region R (that is, a region without holes) then, if C is any closed path within R,

$$\oint_C f(z)\,dz = 0, \tag{5.104}$$

where the symbol $\oint_C$ denotes integration around a simple (non-intersecting) closed curve C (see Figure 5.7). To prove this we refer to

Figure 5.7

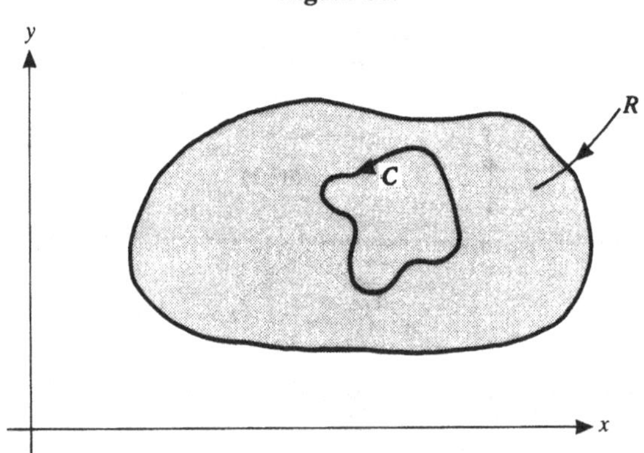

(5.84) which states that

$$\oint_C f(z)\,dz = \oint_C (u\,dx - v\,dy) + \mathrm{i}\oint_C (v\,dx + u\,dy). \qquad (5.105)$$

Now if, in general, $P(x, y)$ and $Q(x, y)$ are two continuous functions with continuous first partial derivatives (at least), then

$$\oint_C (P\,dx + Q\,dy) = -\iint_D \left(\frac{\partial P}{\partial y} - \frac{\partial Q}{\partial x}\right) dx\,dy, \qquad (5.106)$$

where D is the region enclosed by the curve C. This result (called Green's Theorem) may be used to transform the right-hand side of (5.105). Accordingly, we find

$$\oint_C f(z)\,dz = -\iint_D \left(\frac{\partial v}{\partial x} + \frac{\partial u}{\partial y}\right) dx\,dy + \mathrm{i}\iint_D \left(\frac{\partial u}{\partial x} - \frac{\partial v}{\partial y}\right) dx\,dy. \qquad (5.107)$$

Now, since $f(z)$ is analytic, the Cauchy–Riemann equations are satisfied and each integrand on the right-hand side of (5.107) is zero. Hence Cauchy's Theorem (5.104) holds as stated.

Example 8 Evaluate

$$I = \oint_C z\mathrm{e}^z\,dz, \qquad (5.108)$$

where C is a unit circle, centred at the origin.

Since $z\mathrm{e}^z$ is an analytic function (as may be checked by verifying that it satisfies the Cauchy–Riemann equations), $I = 0$ by Cauchy's Theorem. ◢

Example 9 Consider

$$I_1 = \oint_{C_1} \frac{dz}{z-2}, \qquad (5.109)$$

where C_1 is the circle $|z| = 1$ (see Figure 5.8).

Since $1/(z-2)$ is analytic everywhere except at $z = 2$ (which is outside the circle $|z| = 1$), $I_1 = 0$ by Cauchy's Theorem.

Now consider

$$I_2 = \oint_{C_2} \frac{dz}{z-2}, \qquad (5.110)$$

where C_2 is a circle, radius ρ, centred at $z = 2$ (see Figure 5.9).

Writing

$$z = 2 + \rho e^{i\theta}, \tag{5.111}$$

where $0 \leqslant \theta \leqslant 2\pi$, we find using $dz = \rho i e^{i\theta} \, d\theta$,

$$I_2 = \oint_{C_2} \frac{\rho i e^{i\theta} \, d\theta}{\rho e^{i\theta}} = \int_0^{2\pi} i \, d\theta = 2\pi i. \tag{5.112}$$

The non-zero result arises because $1/(z-2)$ is not analytic at $z = 2$ and this point lies *inside* the circle C_2, thus violating the conditions of Cauchy's Theorem. ◢

Example 10 The integral

$$I = \oint z^n \, dz, \tag{5.113}$$

where n is a positive integer, is zero by Cauchy's Theorem since z^n is analytic in the whole finite plane. Consider the circle, radius a, with equation $|z| = a$. Then I can be evaluated directly using polar

Figure 5.8

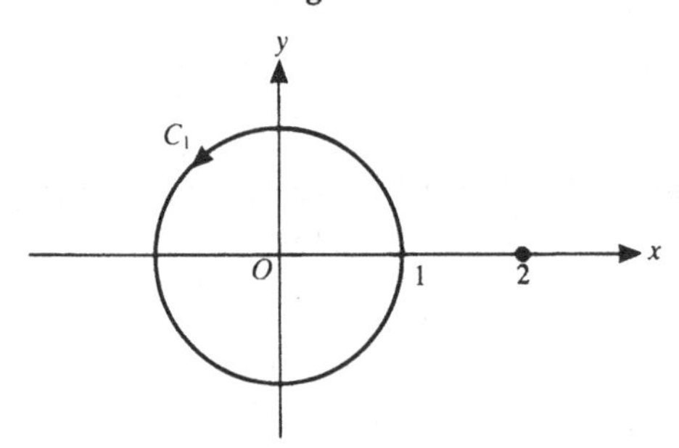

Figure 5.9

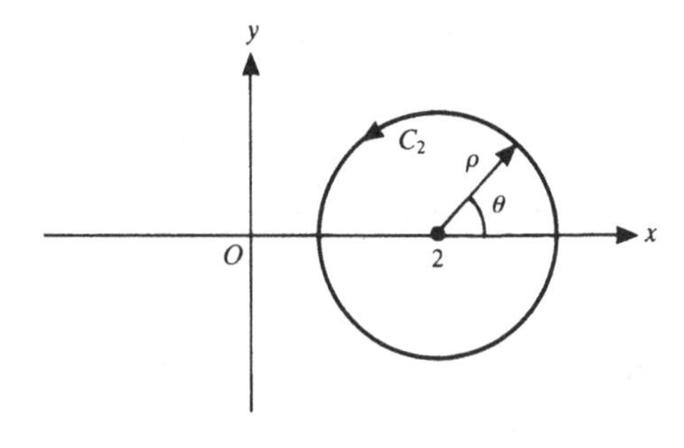

coordinates (as in Example 9). Putting $z = ae^{i\theta}$, we have

$$I = \oint a^n e^{in\theta} a e^{i\theta} i \, d\theta = a^{n+1} i \int_0^{2\pi} e^{i(n+1)\theta} \, d\theta$$

$$= a^{n+1} i \int_0^{2\pi} [\cos(n+1)\theta + i \sin(n+1)\theta] \, d\theta = 0, \quad (5.114)$$

since both the cosine and sine terms integrate to zero. ◢

We now discuss an important result which follows from Cauchy's Theorem. Consider two closed curves C_1 and C_2 in the complex plane and a function $f(z)$ which is analytic on the curves and in the region between them. We form a single closed curve by breaking C_1 and C_2 and joining them by two straight lines AB and EF (see Figure 5.10). Then, by Cauchy's Theorem,

$$\int_{C_1} + \int_A^B + \int_{C_2} + \int_E^F f(z) \, dz = 0, \quad (5.115)$$

since $f(z)$ is analytic on and inside the closed curve where C_1 is described anticlockwise and C_2 clockwise. Now the two straight lines can be made as close as we please and are integrated along in opposite directions. In the limit where these lines coincide their contributions to (5.115) cancel and we have

$$\oint_{C_1} f(z) \, dz + \oint_{C_2} f(z) \, dz = 0. \quad (5.116)$$

Hence

$$\oint_{C_1} f(z) \, dz = -\oint_{C_2} f(z) \, dz = \oint_{C_2} f(z) \, dz. \quad (5.117)$$

Figure 5.10

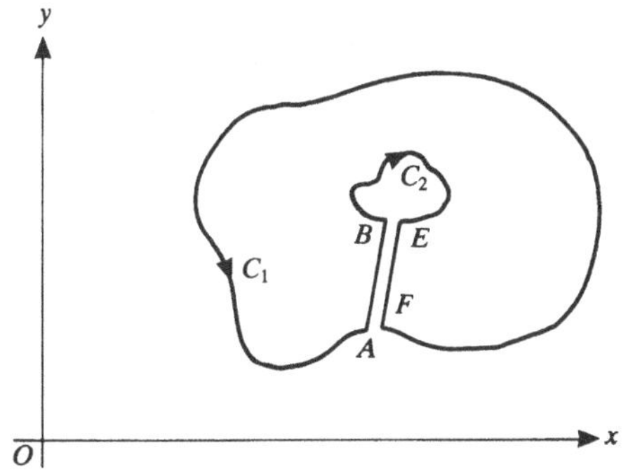

The integral of $f(z)$ around any closed path between C_1 and C_2 is therefore independent of the path.

5.8 Cauchy Integral Formula (Second Integral Theorem)

Suppose we now have a function of the form

$$g(z)=f(z)/(z-z_0), \tag{5.118}$$

where $f(z)$ is assumed to be analytic within a region R and z_0 lies in R. Assuming $f(z_0)\neq 0$, then $g(z)$ is not analytic at $z=z_0$. An example of this type of function was considered in Example 9 where we integrated the function $1/(z-2)$ (which is of the form (5.118) with $f(z)=1$ and $z_0=2$). The Second Cauchy Integral Theorem states that

$$f(z_0)=\frac{1}{2\pi \mathrm{i}}\oint_C \frac{f(z)}{z-z_0}\,dz, \tag{5.119}$$

where C is any closed curve which lies in R and includes the point z_0. To prove this, we surround the point z_0 by a small circle C_1, radius a. By the argument at the end of Section 5.7, we have

$$\oint_C \frac{f(z)}{z-z_0}\,dz=\oint_{C_1}\frac{f(z)}{z-z_0}\,dz, \tag{5.120}$$

since the whole function $f(z)/(z-z_0)$ is analytic between C and C_1. Now let $z=z_0+a\mathrm{e}^{\mathrm{i}\theta}$ describe the circle C_1 so that on the right-hand side of (5.120), $dz=\mathrm{i}a\mathrm{e}^{\mathrm{i}\theta}\,d\theta$ and $z-z_0=a\mathrm{e}^{\mathrm{i}\theta}$. Then

$$\oint_{C_1}\frac{f(z)}{z-z_0}\,dz=\int_0^{2\pi} f(z_0+a\mathrm{e}^{\mathrm{i}\theta})\frac{\mathrm{i}a\mathrm{e}^{\mathrm{i}\theta}}{a\mathrm{e}^{\mathrm{i}\theta}}\,d\theta, \tag{5.121}$$

which, as the small circle C_1 shrinks to zero, becomes

$$f(z_0)\int_0^{2\pi}\mathrm{i}\,d\theta=2\pi\mathrm{i}f(z_0). \tag{5.122}$$

We have used the fact that, since $f(z)$ is a continuous function, $f(z)\to f(z_0)$ as $z\to z_0$. Hence, using (5.120) and (5.122), we have proved the result (5.119). The remarkable property of analytic functions which follows from this result is that the value of $f(z)$ at any point inside a closed contour is completely determined by its values on the contour.

Example 11 Evaluate

$$I=\oint_C \frac{\mathrm{e}^z}{z-1}\,dz, \tag{5.123}$$

where C is the circle $|z-2|=2$.

Clearly the singular point of the integrand is at $z = 1$ which lies within C (see Figure 5.11). Furthermore $f(z) = e^z$ is analytic for any finite point z in the plane. Accordingly, by the Cauchy Integral Formula,

$$I = \oint_C \frac{e^z}{z-1}\,dz = 2\pi i f(z_0 = 1) = 2\pi i e. \quad \blacktriangle \tag{5.124}$$

Example 12 Evaluate

$$I = \oint_C \frac{2z-1}{(z+1)(z-2)}\,dz, \tag{5.125}$$

where (i) $C = C_1$ is the circle $|z| = 4$ and (ii) $C = C_2$ is the circle $|z| = \frac{3}{2}$ (see Figure 5.12).

Figure 5.11

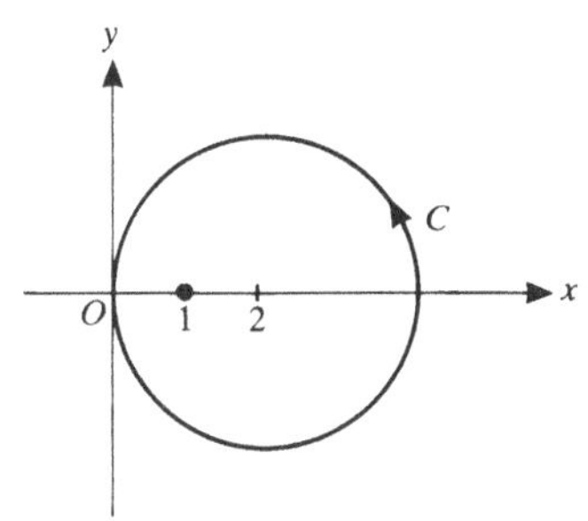

Figure 5.12

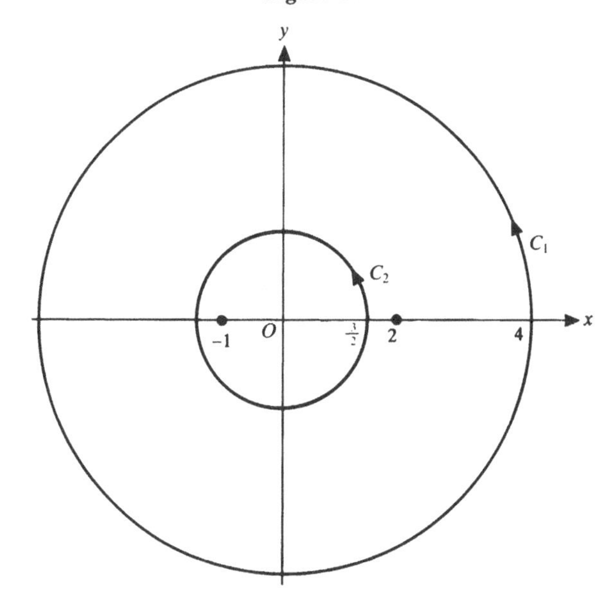

We first use partial fractions to write the integrand as

$$\frac{2z-1}{(z+1)(z-2)} = \frac{1}{z+1} + \frac{1}{z-2}. \tag{5.126}$$

Then

$$\oint_C \frac{2z-1}{(z+1)(z-2)}\,dz = \oint_C \frac{dz}{z+1} + \oint_C \frac{dz}{z-2}. \tag{5.127}$$

(i) For C_1, both singular points $z=-1$ and $z=2$ lie within C_1. Hence, by the Cauchy Integral Formula,

$$\oint_{C_1} \frac{dz}{z+1} = 2\pi\mathrm{i}\,.\,1, \quad \oint_{C_2} \frac{dz}{z-2} = 2\pi\mathrm{i}\,.\,1. \tag{5.128}$$

Hence

$$I = 2\pi\mathrm{i}(1+1) = 4\pi\mathrm{i}. \tag{5.129}$$

(ii) For C_2, only the singular point $z=-1$ lies within C_2. Hence the first integral is

$$\oint_{C_2} \frac{dz}{z+1} = 2\pi\mathrm{i}\,.\,1. \tag{5.130}$$

Since $1/(z-2)$ is analytic within C_2, then by Cauchy's Theorem

$$\oint_{C_2} \frac{dz}{z-2} = 0. \tag{5.131}$$

Hence

$$I = 2\pi\mathrm{i} + 0 = 2\pi\mathrm{i}. \tag{5.132}$$

5.9 Derivatives of an analytic function

We have from (5.46) with $\delta z = h$

$$f'(z_0) = \lim_{h\to 0} \frac{f(z_0+h) - f(z_0)}{h}. \tag{5.133}$$

Now, by (5.119),

$$f(z_0) = \frac{1}{2\pi\mathrm{i}} \oint_C \frac{f(z)}{z-z_0}\,dz \tag{5.134}$$

and

$$f(z_0+h) = \frac{1}{2\pi\mathrm{i}} \oint_C \frac{f(z)}{z-z_0-h}\,dz. \tag{5.135}$$

Substituting (5.134) and (5.135) into (5.133), we have

$$f'(z_0) = \lim_{h\to 0} \frac{1}{2\pi i} \oint_C \left(\frac{1}{z - z_0 - h} - \frac{1}{z - z_0} \right) \frac{f(z)}{h} dz \tag{5.136}$$

$$= \lim_{h\to 0} \frac{1}{2\pi i} \oint_C \frac{f(z)}{(z - z_0 - h)(z - z_0)} dz. \tag{5.137}$$

Reversing the order of the limit and the integration, we have

$$f'(z_0) = \frac{1}{2\pi i} \oint_C \frac{f(z)}{(z - z_0)^2} dz. \tag{5.138}$$

In a similar way, we can obtain the general result

$$f^{(n)}(z_0) = \frac{n!}{2\pi i} \oint_C \frac{f(z)}{(z - z_0)^{n+1}} dz, \tag{5.139}$$

where n is a positive integer and $f^{(n)}$ denotes the nth derivative. Since the analytic function $f(z)$ possesses derivatives $f'(z_0), f''(z_0), \ldots,$ these derivatives must also be analytic functions. Further, if $f(z) = u + iv$ then the existence of f'' implies that f' is differentiable and hence that $\partial u/\partial x$ and $\partial v/\partial y$ may be differentiated. From the Cauchy–Riemann equations (5.56) we then have

$$\partial^2 u/\partial x^2 = \partial^2 v/\partial x\, \partial y = -\partial^2 u/\partial y^2 \tag{5.140}$$

so that

$$\nabla^2 u = \frac{\partial^2 u}{\partial x^2} + \frac{\partial^2 u}{\partial y^2} = 0. \tag{5.141}$$

Similarly

$$\nabla^2 v = \frac{\partial^2 v}{\partial x^2} + \frac{\partial^2 v}{\partial y^2} = 0. \tag{5.142}$$

Example 13 Evaluate

$$I = \oint_C \frac{e^{2z}}{(z - 1)^2} dz, \tag{5.143}$$

where C is the circle $|z| = 2$.

Here the singular point of the integrand is at $z = 1$, and $f(z) = e^{2z}$. Comparing with (5.138), we see that

$$\oint_C \frac{e^{2z}}{(z - 1)^2} dz = 2\pi i f'(z_0 = 1) = 2\pi i \,.\, 2e^2 = 4\pi i e^2. \quad ◢ \tag{5.144}$$

5.10 Taylor and Laurent series

We now state, without proof, two important theorems.

1. Taylor's Theorem

If $f(z)$ is analytic within and on a circle C of radius R and centre z_0, then

$$f(z)=\sum_{n=0}^{\infty}\frac{f^{(n)}(z_0)}{n!}(z-z_0)^n. \tag{5.145}$$

This series converges within the circle, centre z_0, of radius R equal to the distance from z_0 to the nearest singularity of $f(z)$. In general, $f^{(n)}(z_0)$ can be found from (5.139).

Example 14 To find the Taylor series for $f(z)=\sin z$ about $z_0=0$, we first evaluate the derivatives

$$f'(z)=\cos z,\quad f'(0)=1, \tag{5.146}$$

$$f''(z)=-\sin z,\quad f''(0)=0, \tag{5.147}$$

$$f'''(z)=-\cos z,\quad f'''(0)=-1, \tag{5.148}$$

and so on. Now from (5.145)–(5.148),

$$\sin z=\sum_{n=0}^{\infty}\frac{f^{(n)}(0)}{n!}z^n=z-\frac{z^3}{3!}+\frac{z^5}{5!}-\ldots+\frac{(-1)^{n+1}z^{2n+1}}{(2n+1)!}+\ldots, \tag{5.149}$$

which is the familiar series for the sine function. ◢

Example 15 The function $1/z^2$ is not analytic in a region containing the origin $z=0$. However, inside the circle $|z-1|=1$ (see Figure 5.13) the function is analytic and may therefore be expanded as a

Figure 5.13

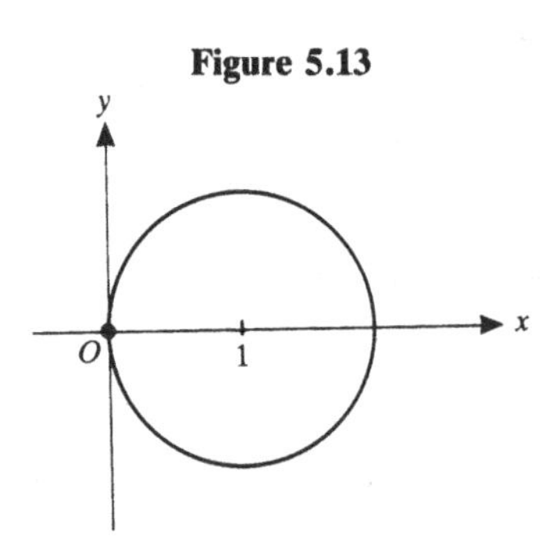

Taylor series about $z = 1$. By differentiation,

$$f^{(n)}(z) = (-1)^n \frac{(n+1)!}{z^{n+2}} \tag{5.150}$$

and hence

$$f^{(n)}(1) = (-1)^n (n+1)! \tag{5.151}$$

Accordingly the Taylor series (5.145) gives

$$\frac{1}{z^2} = \sum_{n=0}^{\infty} (-1)^n \frac{(n+1)!}{n!} (z-1)^n = \sum_{n=0}^{\infty} (-1)^n (n+1)(z-1)^n. \tag{5.152}$$

For convergence, $|z-1|<1$ since the distance from the centre of expansion ($z_0 = 1$) to the nearest singularity of $1/z^2$ (at $z = 0$) is unity. ◢

Before stating the second theorem, we first explain what is meant by an isolated singularity. Consider a single-valued function $f(z)$ with a singularity at $z = z_0$. We surround the point $z = z_0$ by a circle centred on z_0. If the function is analytic everywhere within some sufficiently small circle, except at the single point $z = z_0$, then z_0 is termed an isolated singularity. For example, $f(z) = 1/z^2$ has an isolated singularity at $z = 0$ since within an arbitrarily small circle enclosing this point the function is analytic except at $z = 0$. However, consider the function $f(z) = \operatorname{cosec}(1/z) = 1/\sin(1/z)$. This function has singularities at $z = 0$ and at $z = 1/n\pi$, $n = \pm 1, \pm 2, \ldots$. If we move along the x-axis towards the origin, we encounter an ever-increasing density of singularities, that is, the singularities accumulate onto the point $z = 0$. Each of the singularities at $1/n\pi$ is isolated since an infinitesimally small circle can be placed around each one so that the circle contains only one singular point. However, the singularity at $z = 0$ is not isolated since any circle, however small, centred on $z = 0$ will contain an infinite number of singular points. For multi-valued functions such as $\sqrt{z}$ and $\ln z$, the concept of an isolated singularity is not applicable.

2. Laurent's Theorem

This is an extension of the preceding theorem and states the following: If $f(z)$ is analytic inside and on the boundaries of the circular annulus defined by $|z - z_0| = R_1$ and $|z - z_0| = R_2$ with $R_2 < R_1$ (see Figure 5.14), then $f(z)$ can be represented by the Laurent series

$$f(z) = \sum_{n=0}^{\infty} a_n (z - z_0)^n + \sum_{n=1}^{\infty} \frac{a_{-n}}{(z - z_0)^n}, \tag{5.153}$$

where

$$a_n = \frac{1}{2\pi i} \oint_{C_1} \frac{f(z)}{(z - z_0)^{n+1}} dz \tag{5.154}$$

and

$$a_{-n} = \frac{1}{2\pi i} \oint_{C_2} \frac{f(z)}{(z - z_0)^{-n+1}} dz. \tag{5.155}$$

In the case when z_0 is an isolated singularity of $f(z)$, the series (5.153) consists of a Taylor series (the a_n series, convergent for $|z - z_0| < R_1$), and the a_{-n} series which contains terms which are singular at z_0. This latter series, called the principal part of the Laurent series and denoted by $P(z)$, converges for all $|z - z_0| > 0$ (that is, R_2 may be taken to be zero).

Example 16 Consider

$$f(z) = e^z/z^4, \tag{5.156}$$

which has an isolated singularity at $z = 0$. Then since

$$e^z = 1 + z + \frac{z^2}{2!} + \ldots + \frac{z^n}{n!} + \ldots, \tag{5.157}$$

we have, for all $z \neq 0$,

$$f(z) = \frac{1}{z^4} + \frac{1}{z^3} + \frac{1}{2!\, z^2} + \frac{1}{3!\, z} + \frac{1}{4!} + \frac{z}{5!} + \ldots. \tag{5.158}$$

This is the Laurent series expansion (5.153) about $z = 0$. The principal part of the series is

$$P(z) = \frac{1}{z^4} + \frac{1}{z^3} + \frac{1}{2!\, z^2} + \frac{1}{3!\, z}, \tag{5.159}$$

which is a finite series. ◢

Figure 5.14

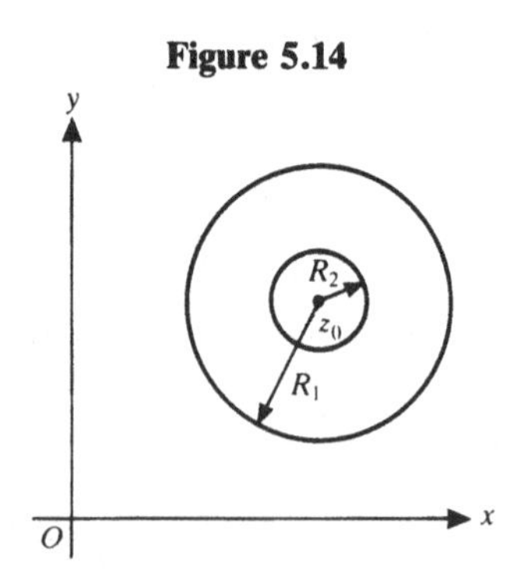

Example 17 Not all principal parts of Laurent series are finite series. Consider, for example,

$$f(z) = z\mathrm{e}^{1/z} = z\left(1 + \frac{1}{z} + \frac{1}{2!\,z^2} + \frac{1}{3!\,z^3} + \ldots\right) \tag{5.160}$$

$$= z + 1 + \frac{1}{2!\,z} + \frac{1}{3!\,z^2} + \ldots, \tag{5.161}$$

for all $z \neq 0$. This function has

$$P(z) = \frac{1}{2!\,z} + \frac{1}{3!\,z^2} + \ldots + \frac{1}{n!\,z^{n-1}} + \ldots, \tag{5.162}$$

which is an infinite series. ◢

Example 18 Both the previous examples had singularities at $z = 0$. Suppose now

$$f(z) = \mathrm{e}^z/(z-1)^2. \tag{5.163}$$

Then $f(z)$ is singular at $z = 1$. Hence let $z - 1 = u$, say, so that

$$\frac{\mathrm{e}^z}{(z-1)^2} = \frac{\mathrm{e}^{1+u}}{u^2} = \frac{\mathrm{e}}{u^2}\left(1 + u + \frac{u^2}{2!} + \frac{u^3}{3!} + \ldots\right), \tag{5.164}$$

giving

$$f(z) = \mathrm{e}\left[\frac{1}{(z-1)^2} + \frac{1}{z-1} + \frac{1}{2!} + \frac{(z-1)}{3!} + \ldots\right], \tag{5.165}$$

which is the Laurent series expansion of (5.163) about $z = 1$. The principal part is

$$P(z) = \mathrm{e}\left[\frac{1}{(z-1)^2} + \frac{1}{(z-1)}\right]. \tag{5.166}$$

◢

Example 19 Consider

$$f(z) = \frac{1}{z(z-1)} = \frac{1}{z-1} - \frac{1}{z}. \tag{5.167}$$

The singularities of $f(z)$ are at $z = 0$ and $z = 1$. Hence there are two regions in which the function is analytic and for which we can write a Laurent series in powers of z: (i) $0 < |z| < 1$ and (ii) $1 < |z|$. We consider these regions separately.

(i) For $0 < |z| < 1$,

$$f(z) = -\frac{1}{z} - (1-z)^{-1}. \tag{5.168}$$

Now, since $0<|z|<1$, the second term can be expanded by the Binomial Theorem to give

$$f(z)=-\frac{1}{z}-(1+z+z^2+z^3+\ldots). \tag{5.169}$$

(ii) For $|z|>1$,

$$f(z)=-\frac{1}{z}+\frac{1}{z}\left(1-\frac{1}{z}\right)^{-1}. \tag{5.170}$$

Again, since $|z|>1$, the second term can be expanded by the Binomial Theorem giving

$$f(z)=-\frac{1}{z}+\frac{1}{z}\left(1+\frac{1}{z}+\frac{1}{z^2}+\frac{1}{z^3}+\ldots\right) \tag{5.171}$$

$$=\frac{1}{z^2}+\frac{1}{z^3}+\frac{1}{z^4}+\ldots. \tag{5.172}$$

In the above we have chosen $z_0=0$ as the centre of expansion. If we choose instead $z_0=1$, then we obtain two series in powers of $z-1$, one for $0<|z-1|<1$ and one for $|z-1|>1$. The first of these is obtained by putting $z-1=u$ so that

$$\frac{1}{z(z-1)}=\frac{1}{u}-\frac{1}{1+u}=\frac{1}{u}-(1-u+u^2-\ldots), \tag{5.173}$$

since $|u|<1$, giving

$$f(z)=\frac{1}{z-1}-1+(z-1)-(z-1)^2+\ldots, \tag{5.174}$$

Similarly, we may obtain the series for $|u|>1$. ◢

Example 20 Consider

$$f(z)=\mathrm{Ln}\left(\frac{z^2}{1-z^2}\right) \tag{5.175}$$

for $|z|>1$ (recalling that the capital letter indicates the principal value of the logarithm). Then

$$f(z)=\mathrm{Ln}\left(\frac{1}{1/z^2-1}\right)=\mathrm{Ln}\left(\frac{-1}{1-1/z^2}\right). \tag{5.176}$$

Hence

$$f(z)=\mathrm{Ln}(-1)-\mathrm{Ln}(1-1/z^2). \tag{5.177}$$

The principal value of $\mathrm{Ln}(-1)$ is $\mathrm{i}\pi$ while the second term can be expanded using the series for $\mathrm{Ln}(1-1/z^2)$ since $|1/z^2|<1$. We find

$$f(z) = \mathrm{i}\pi - \left(-\frac{1}{z^2} - \frac{1}{2z^4} - \frac{1}{3z^6} - \dots\right)$$

$$= \mathrm{i}\pi + \frac{1}{z^2} + \frac{1}{2z^4} + \frac{1}{3z^6} + \dots \quad ◢ \tag{5.178}$$

5.11 Singularities and residues

We have seen in Section 5.10 that if $f(z)$ has an isolated singularity at $z = z_0$ then in the neighbourhood of this point $f(z)$ can be represented by the Laurent series (5.153). The principal part of the Laurent series (the terms with negative powers of $z - z_0$) may contain either a finite number of terms (as in Example 16 above) or an infinite number of terms (as in Example 17 above). We have the following definitions in each case:

Finite number of terms. In this case the singularity at z_0 is called a pole of order m, where m is the highest negative power of $z - z_0$. Hence, in Example 16, $f(z)$ has a pole of order 4 at $z = 0$ whilst in Example 18, $f(z)$ has a pole of order 2 at $z = 1$. Poles of order 1 are called simple poles. Hence, in Example 19, $f(z)$ has a simple pole at $z = 0$ (see the Laurent series (5.169) about $z = 0$ for $0 < |z| < 1$) and a simple pole at $z = 1$ (see the Laurent series (5.174) about $z = 1$ for $0 < |z - 1| < 1$).

Infinite number of terms. When the Laurent series in the neighbourhood of a singular point contains infinitely many terms with negative powers of $z - z_0$, the point $z = z_0$ is termed an essential singularity. For example

$$f(z) = \mathrm{e}^{1/z} = 1 + \frac{1}{z} + \frac{1}{2!\, z^2} + \dots \tag{5.179}$$

has an infinity of terms in negative powers of z. Hence $z = 0$ is an essential singularity of $\mathrm{e}^{1/z}$. Similarly

$$f(z) = \sin\left(\frac{1}{z}\right) = \frac{1}{z} - \frac{1}{3!\, z^3} + \frac{1}{5!\, z^5} - \dots \tag{5.180}$$

has an essential singularity at $z = 0$.

An important number, as we shall see in Section 5.12, is the coefficient a_{-1} in the Laurent series. This coefficient is called the residue of the function $f(z)$ at $z = z_0$ and we now show how it may be calculated in different cases.

1. *Simple poles*

Clearly if we have a simple pole at $z = z_0$, then the Laurent series (5.153) will have the form

$$f(z) = \frac{a_{-1}}{z - z_0} + \sum_{n=0}^{\infty} a_n(z - z_0)^n. \tag{5.181}$$

By multiplying through by $z - z_0$, we have

$$(z - z_0)f(z) = a_{-1} + a_0(z - z_0) + a_1(z - z_0)^2 + \ldots. \tag{5.182}$$

Hence taking the limit as $z \to z_0$, we find

$$\lim_{z \to z_0} [(z - z_0)f(z)] = a_{-1}, \tag{5.183}$$

which enables the residue to be calculated.

Example 21 To find the residues at $z = 0$ and $z = 1$ of

$$f(z) = 1/z(z - 1). \tag{5.184}$$

Method 1 In Example 19 above we showed that the Laurent series of $f(z)$ in the neighbourhood of $z = 0$ is (see (5.169))

$$f(z) = -\frac{1}{z} - 1 - z - z^2 - \ldots. \tag{5.185}$$

The coefficient of $1/z$ is by definition the residue at $z = 0$ and is therefore -1. Further we showed (see (5.174)) that the Laurent series of $f(z)$ in the neighbourhood of $z = 1$ is

$$f(z) = \frac{1}{z - 1} - 1 + (z - 1) - (z - 1)^2 + \ldots. \tag{5.186}$$

Hence the residue at $z = 1$ (the coefficient of $1/(z - 1)$) is 1.

Method 2 We now use the formula (5.183). For $z = 0$,

$$a_{-1} = \lim_{z \to 0}\left[z \frac{1}{z(z - 1)}\right] = -1, \tag{5.187}$$

as before.

For $z = 1$,

$$a_{-1} = \lim_{z\to 1}\left[(z-1)\frac{1}{z(z-1)}\right] = 1, \tag{5.188}$$

as above. ◢

Example 22 To show that $z = 3\pi/2$ is a simple pole of

$$f(z) = \mathrm{e}^{2z}\tan z, \tag{5.189}$$

and to determine the residue there.

Let $z - 3\pi/2 = u$. Then expressing (5.189) in terms of u we have

$$\begin{aligned} f(u) &= \mathrm{e}^{2(u+3\pi/2)}\tan(u+3\pi/2) \\ &= \mathrm{e}^{(2u+3\pi)}\frac{\sin(u+3\pi/2)}{\cos(u+3\pi/2)} \\ &= -\mathrm{e}^{(2u+3\pi)}\frac{\cos u}{\sin u}. \end{aligned} \tag{5.190}$$

We now expand the functions $\cos u$ and $\sin u$ as power series in u so that

$$\begin{aligned} f(u) &= -\mathrm{e}^{3\pi}\mathrm{e}^{2u}\frac{\left(1-\dfrac{u^2}{2!}+\dfrac{u^4}{4!}-\dots\right)}{\left(u-\dfrac{u^3}{3!}+\dfrac{u^5}{5!}-\dots\right)} \\ &= -\mathrm{e}^{3\pi}\frac{\mathrm{e}^{2u}}{u}\left(1-\frac{u^2}{2!}+\frac{u^4}{4!}-\dots\right)\left(1-\frac{u^2}{3!}+\frac{u^4}{5!}-\dots\right)^{-1}. \end{aligned} \tag{5.191}$$

We may further expand e^{2u} as a power series and use the Binomial Theorem to expand the last factor. Then

$$\begin{aligned} f(u) = -\frac{\mathrm{e}^{3\pi}}{u}&\left[1+2u+\frac{(2u)^2}{2!}+\dots\right]\left(1-\frac{u^2}{2!}+\frac{u^4}{4!}-\dots\right) \\ &\times\left(1+\frac{u^2}{3!}-\dots\right). \end{aligned} \tag{5.192}$$

Since all the expansions involve only positive (or zero) powers of u, the only term with a negative power of u is $-\mathrm{e}^{3\pi}/u$. Hence the principal part of the Laurent series of $f(z)$ in the neighbourhood of $z = 3\pi/2$ is

$$P(z) = -\frac{\mathrm{e}^{3\pi}}{z-3\pi/2}. \tag{5.193}$$

The pole is therefore a simple one. The residue is the coefficient of $1/(z-3\pi/2)$, which is $-e^{3\pi}$.

Alternatively, the residue could be calculated using (5.183). We obtain

$$a_{-1} = \lim_{z\to 3\pi/2} [(z-3\pi/2)e^{2z}\tan z]$$
$$= \lim_{z\to 3\pi/2}\left[(z-3\pi/2)e^{2z}\frac{\sin z}{\cos z}\right]. \tag{5.194}$$

Now both $z-3\pi/2$ and $\cos z$ tend to zero as $z\to 3\pi/2$. Hence

$$a_{-1} = \lim_{z\to 3\pi/2}\left(\frac{z-3\pi/2}{\cos z}\right)\left(\lim_{z\to 3\pi/2} e^{2z}\right)\left(\lim_{z\to 3\pi/2}\sin z\right)$$
$$= \lim_{z\to 3\pi/2}\left(\frac{1}{-\sin z}\right)(e^{3\pi})(-1), \tag{5.195}$$

where we have used L'Hôpital's rule in the first term. We find therefore

$$a_{-1} = \frac{-e^{3\pi}}{-\sin(3\pi/2)} = -e^{3\pi}, \tag{5.196}$$

as above. ◢

Besides the two methods described above, the residue at a simple pole may be calculated by another method. Suppose

$$f(z) = \frac{p(z)}{q(z)} \tag{5.197}$$

and that $z=z_0$ is a *simple* zero of $q(z)$ (that is, $q(z)$ behaves like $z-z_0$ as $z\to z_0$). Then $z=z_0$ is a simple pole of $f(z)$ provided $p(z_0)\neq 0$ and that $p(z)$ is not singular at $z=z_0$. Using (5.183) we can now write

$$a_{-1} = \lim_{z\to z_0}\left[(z-z_0)\frac{p(z)}{q(z)}\right] = \lim_{z\to z_0}\left[\frac{p(z)}{q(z)/(z-z_0)}\right]$$
$$= \lim_{z\to z_0}\left\{\frac{p(z)}{\left[\dfrac{q(z)-q(z_0)}{z-z_0}\right]}\right\}, \tag{5.198}$$

since $q(z_0)=0$ by definition. Hence

$$a_{-1} = \lim_{z\to z_0} p(z)\frac{1}{\lim_{z\to z_0}\left[\dfrac{q(z)-q(z_0)}{z-z_0}\right]}$$
$$= p(z_0)\frac{1}{q'(z_0)}, \tag{5.199}$$

using the limit definition of the derivative of $q(z)$ at $z = z_0$. For a simple pole, therefore,

$$a_{-1} = p(z_0)/q'(z_0). \tag{5.200}$$

Example 23 Consider

$$f(z) = \frac{1}{1 - z^3}. \tag{5.201}$$

The pole at $z = 1$ is simple (since $z = 1$ is a simple zero of $1 - z^3 = (z-1)(-1-z-z^2)$). Hence, since $p(z) = 1$ and $q(z) = 1 - z^3$,

$$a_{-1} = \frac{1}{(-3z^2)_{z=1}} = -\tfrac{1}{3}. \quad ◢ \tag{5.202}$$

2. Poles of order m

The Laurent series (5.153) for a function $f(z)$ with a pole of order m at $z - z_0$ has the form

$$f(z) = \sum_{n=0}^{\infty} a_n (z - z_0)^n + \frac{a_{-1}}{(z - z_0)} + \frac{a_{-2}}{(z - z_0)^2} + \ldots + \frac{a_{-m}}{(z - z_0)^m}. \tag{5.203}$$

Now multiplying (5.203) by $(z - z_0)^m$ we have

$$(z - z_0)^m f(z) = \sum_{n=0}^{\infty} a_n (z - z_0)^{m+n} + a_{-1}(z - z_0)^{m-1} + a_{-2}(z - z_0)^{m-2} + \ldots + a_{-m}. \tag{5.204}$$

To find a_{-1}, we differentiate (5.204) $m - 1$ times to obtain

$$\frac{d^{m-1}}{dz^{m-1}}[(z - z_0)^m f(z)] = (m-1)!\, a_{-1} + m!\, a_0(z - z_0) + \ldots. \tag{5.205}$$

Hence

$$a_{-1} = \lim_{z \to z_0} \frac{1}{(m-1)!} \frac{d^{m-1}}{dz^{m-1}}[(z - z_0)^m f(z)]. \tag{5.206}$$

When $m = 1$ we obtain the result (5.183) for a simple pole.

Example 24 Consider

$$f(z) = \left(\frac{z}{z-1}\right)^2. \tag{5.207}$$

Then $z = 1$ is a double pole (pole of order 2). We can obtain the residue at this point by finding the Laurent series, or by using (5.206).

Method 1 Let $z-1=u$. Then

$$\left(\frac{z}{z-1}\right)^2=\frac{(u+1)^2}{u^2}=\frac{1+2u+u^2}{u^2}=\frac{1}{u^2}+\frac{2}{u}+1. \tag{5.208}$$

Hence the Laurent series is

$$f(z)=\frac{1}{(z-1)^2}+\frac{2}{(z-1)}+1. \tag{5.209}$$

The coefficient of $1/(z-1)$ is 2 which is therefore the residue at $z=1$.

Method 2 Using (5.206) with $m=2$,

$$a_{-1}=\lim_{z\to 1}\left\{\frac{1}{1!}\frac{d}{dz}\left[(z-1)^2\left(\frac{z}{z-1}\right)^2\right]\right\} \tag{5.210}$$

$$=\lim_{z\to 1}\frac{d}{dz}(z^2)=2, \tag{5.211}$$

as in method 1. ◢

Example 25 Consider

$$f(z)=\frac{1}{(e^z-1)^2}. \tag{5.212}$$

Then $f(z)$ has a pole of order 2 at $z=0$. Using the expansion of e^z, we find

$$f(z)=\frac{1}{(z+z^2/2!+\dots)^2}=\frac{1}{z^2}\left(1+\frac{z}{2}+\dots\right)^{-2}=\frac{1}{z^2}-\frac{1}{z}+\dots. \tag{5.213}$$

From the Laurent series (5.213), we see that the residue of $f(z)$ at $z=0$ is -1. Alternatively, using (5.206), we have

$$a_{-1}=\lim_{z\to 0}\left\{\frac{d}{dz}\left[z^2\frac{1}{(e^z-1)^2}\right]\right\} \tag{5.214}$$

$$=\lim_{z\to 0}\left[\frac{2z(e^z-1)-2z^2e^z}{(e^z-1)^3}\right]. \tag{5.215}$$

Both the numerator and the denominator of this expression tend to zero as $z\to 0$. The value of the limit may be obtained by expanding e^z

as follows:

$$a_{-1} = \lim_{z\to 0}\left[\frac{2z\left(z+\frac{z^2}{2!}+\frac{z^3}{3!}+\ldots\right)-2z^2\left(1+z+\frac{z^2}{2!}+\ldots\right)}{\left(z+\frac{z^2}{2!}+\frac{z^3}{3!}+\ldots\right)^3}\right] \quad (5.216)$$

$$= \lim_{z\to 0}\left[\frac{-z^3-\frac{2}{3}z^4+\ldots}{z^3\left(1+\frac{z}{2!}+\ldots\right)^3}\right] \quad (5.217)$$

$$= \lim_{z\to 0}\left[\frac{-1-\frac{2}{3}z+\ldots}{\left(1+\frac{z}{2!}+\ldots\right)^3}\right] = -1. \quad (5.218)$$

Hence the residue of $f(z)$ at the double pole $z=0$ is -1, as before. ◢

5.12 Cauchy Residue Theorem

We have seen that, in the neighbourhood of a singularity z_0, $f(z)$ may be represented by the Laurent series

$$f(z) = \sum_{n=1}^{m}\frac{a_{-n}}{(z-z_0)^n} + \sum_{n=0}^{\infty} a_n(z-z_0)^n, \quad (5.219)$$

where the singularity is termed a pole of order m for any finite m, and an essential singularity for m infinite. Suppose we require

$$I = \oint_C f(z)\,dz, \quad (5.220)$$

where C is some closed curve surrounding z_0. We showed in Section 5.7 that the integral around C is equal to the integral around a circle γ of radius ρ, centre z_0. This was done by cutting C and inserting two parallel straight lines of equal length joining C to γ, the contributions from these lines (the cut) cancelling as the lines are brought into coincidence. Hence

$$\oint_C f(z)\,dz = \oint_\gamma f(z)\,dz. \quad (5.221)$$

Using (5.221) and (5.219) we have

$$\oint_C f(z)\,dz = \oint_\gamma\left[\sum_{n=1}^{m}\frac{a_{-n}}{(z-z_0)^n} + \sum_{n=0}^{\infty} a_n(z-z_0)^n\right]dz. \quad (5.222)$$

The integral of each term in the second summation is zero by Cauchy's Theorem since each term is an analytic function. Now consider a typical term in the first summation, say,

$$\oint_{\gamma} \frac{a_{-n}}{(z-z_0)^n}\, dz. \tag{5.223}$$

Since γ is a circle of radius ρ (say), then on γ, $z = z_0 + \rho e^{i\theta}$, and (5.223) can be evaluated directly:

$$\oint_{\gamma} \frac{a_{-n}}{(z-z_0)^n}\, dz = a_{-n} \int_0^{2\pi} \frac{\rho e^{i\theta} i\, d\theta}{\rho^n e^{in\theta}} = \frac{ia_{-n}}{\rho^{n-1}} \int_0^{2\pi} e^{-i(n-1)\theta}\, d\theta. \tag{5.224}$$

If $n \neq 1$, this integral is simply

$$\frac{ia_{-n}}{\rho^{n-1}} \int_0^{2\pi} \{\cos[(n-1)\theta] - i \sin[(n-1)\theta]\}\, d\theta = 0. \tag{5.225}$$

If $n = 1$, then

$$\oint_{\gamma} \frac{a_{-1}}{(z-z_0)}\, dz = ia_{-1} \int_0^{2\pi} d\theta = 2\pi i a_{-1}. \tag{5.226}$$

Accordingly,

$$\oint_C f(z)\, dz = 2\pi i a_{-1}. \tag{5.227}$$

From Section 5.11, a_{-1} is the residue of $f(z)$ at the pole of order m at $z = z_0$. Since we know how to calculate a_{-1} for a given function $f(z)$, it follows that we can now calculate $\oint_C f(z)\, dz$ around any closed contour C surrounding a singularity at z_0. The result (5.227), known as the Cauchy Residue Theorem, may be readily extended to the case where $f(z)$ has a number of poles $z_0, z_1, z_2, \ldots, z_k$ within some closed contour C (the case of three singularities is shown in Figure 5.15).

Figure 5.15

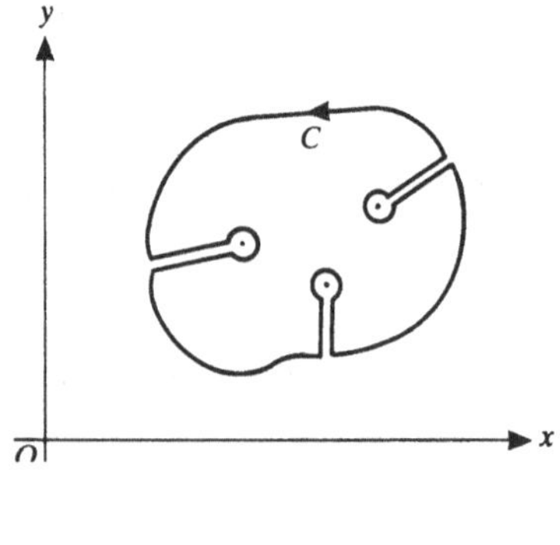

Forming the cuts in the usual way, we see that the integral around C is just the sum of the integrals around circles centred on each singularity. Hence

$$\oint_C f(z)\, dz = 2\pi \mathrm{i}(a_{-1} + b_{-1} + c_{-1} + \ldots), \tag{5.228}$$

where $a_{-1}, b_{-1}, c_{-1}, \ldots$ are the residues at $z_0, z_1, z_2, \ldots$. We therefore have

$$\oint_C f(z)\, dz = 2\pi \mathrm{i} \times (\text{sum of the residues at the poles within } C). \tag{5.229}$$

Example 26 Consider the function

$$f(z) = \mathrm{e}^z / z(z-1). \tag{5.230}$$

Then $f(z)$ has two simple poles, one at $z = 0$ and one at $z = 1$. At $z = 0$, using (5.183), the residue is

$$a_{-1} = \lim_{z\to 0}\left[z \frac{\mathrm{e}^z}{z(z-1)}\right] = -1, \tag{5.231}$$

whilst at $z = 1$, the residue is

$$b_{-1} = \lim_{z\to 1}\left[(z-1)\frac{\mathrm{e}^z}{z(z-1)}\right] = \mathrm{e}. \tag{5.232}$$

We now evaluate the integral of $\mathrm{e}^z/z(z-1)$ around three different closed curves C_1, C_2, and C_3 (see Figure 5.16). Consider first

$$\oint_{C_1} \frac{\mathrm{e}^z}{z(z-1)}\, dz, \tag{5.233}$$

where C_1 is some closed curve not containing a pole. Then the

Figure 5.16

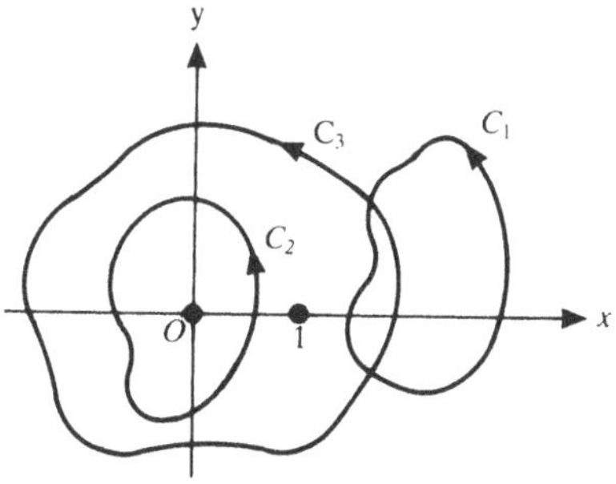

function is analytic inside and on C_1 and hence by Cauchy's Theorem the integral is zero. On the other hand, by the Cauchy Residue Theorem,

$$\oint_{C_2} \frac{e^z}{z(z-1)}\,dz = 2\pi i(-1), \tag{5.234}$$

where C_2 encloses the singular point $z = 0$ (at which the residue is -1) but not the singular point $z = 1$. Finally,

$$\oint_{C_3} \frac{e^z}{z(z-1)}\,dz = 2\pi i(e-1), \tag{5.235}$$

where C_3 encloses both poles at $z = 0$ and $z = 1$ at which the residues are respectively -1 and e. ◢

The Cauchy Residue Theorem enables many real definite integrals to be evaluated without performing any integration. This and other applications are dealt with in the next chapter.

Problems 5

1. If $z = x + iy$, show that, if $a > 0$, $|a^z| = a^x$, where the principal value of a^z is taken.
2. Determine all possible values of

   (i) $\ln(1 - i)$, (ii) $\tan^{-1}(2i)$.

3. Show that the function e^z has no zeros, and locate the zeros of $\cosh z$.
4. Given that

   $$u = x^2 + 4x - y^2 + 2y$$

   is the real part of an analytic function $f(z)$, determine the corresponding imaginary part v and the form of $f(z)$.
5. If $f(z)$ is an analytic function, show that

   $$\nabla^2 |f(z)|^2 = 4\,|f'(z)|^2.$$

   Use this result to find the most general forms of $g(z)$ and $h(z)$ given that $g(z)$ and $h(z)$ are analytic for all z, and that $\phi = |g(z)|^2 + |h(z)|^2$ is harmonic (that is, it satisfies Laplace's equation $\nabla^2\phi = 0$).

6. Evaluate $\int (z^2+1)\,dz$
   (i) along the straight line joining the origin, $z=0$, to $z=1+\mathrm{i}$,
   (ii) along the parabola $x=t$, $y=t^2$ where $0 \leqslant t \leqslant 1$,
   (iii) along the straight line from the origin, $z=0$, to $z=1$, and then along the straight line from $z=1$ to $z=1+\mathrm{i}$.

   Why are the results the same?
7. (i) Evaluate

$$\oint_C \frac{dz}{z-2},$$

where C is (a) the circle $|z|=1$; (b) the circle $|z+\mathrm{i}|=4$.

(ii) By writing $z=\mathrm{e}^{\mathrm{i}\theta}$, evaluate $\int_C \operatorname{Ln} z\,dz$ and $\int_C z^3 \operatorname{Ln} z\,dz$, where C is the curve defined by $-\pi<\theta<\pi$.

8. Evaluate the following integrals using the Cauchy Integral Formula, where C is the circle $|z|=2$:

$$\text{(i)}\ \oint_C \frac{\sin z}{z^2+1}\,dz, \quad \text{(ii)}\ \oint_C \frac{z\mathrm{e}^z}{(z-1)^4}\,dz.$$

9. Evaluate the following integrals using the Cauchy Integral Formula, where C is the boundary of a square the sides of which are defined by $x=\pm 2$, $y=\pm 2$:

$$\text{(i)}\ \oint_C \frac{\mathrm{e}^{-z}}{z-\pi\mathrm{i}/2}\,dz, \quad \text{(ii)}\ \oint_C \frac{\cos z}{z(z^2+9)}\,dz,$$

$$\text{(iii)}\ \oint_C \frac{\tan(z/2)}{(z-\pi/2)^2}\,dz.$$

10. Find the Taylor series about $z=0$ and its radius of convergence for each of the following functions:

$$\text{(i)}\ \mathrm{e}^{z\sin z}, \quad \text{(ii)}\ \sin\left(\frac{1}{1-z}\right), \quad \text{(iii)}\ \operatorname{Ln}(1+\mathrm{e}^z).$$

11. Locate the singularities of the following functions and determine their nature and the appropriate Laurent series about these points:

$$\text{(i)}\ z\sin(1/z), \quad \text{(ii)}\ \frac{\cos(2z)}{z-\pi}, \quad \text{(iii)}\ \frac{\mathrm{e}^z \sin z}{(z-\pi/2)^2}.$$

12. Find the Laurent series of $1/z(z-1)^2$ about
    (i) $z=0$, $0<|z|<1$,
    (ii) $z=0$, $|z|>1$,
    (iii) $z=1$, $0<|z-1|<1$,
    (iv) $z=1$, $|z-1|>1$.

13. Find the residues of each function at the specified poles:

 (i) $\dfrac{1}{1-\mathrm{e}^z}$ at all the poles,

 (ii) $\cot z$ at $z = 3\pi$,

 (iii) $\dfrac{1}{z^3(z-1)^2(z-2)}$ at all the poles.

14. Locate the poles of $z/(1-a\mathrm{e}^{-\mathrm{i}z})$, where $a > 1$ is a real constant, and hence deduce that

$$\oint_C \frac{z}{1-a\mathrm{e}^{-\mathrm{i}z}}\,dz = 0,$$

where C is any closed contour in the upper half-plane.

15. Show that

$$\oint_C \frac{\mathrm{e}^{imz}}{z^2+1}\,dz = \pi\mathrm{e}^{-m},$$

where C is a contour lying in the upper half-plane consisting of the semicircle of radius $R > 1$, centred at the origin, and the portion of the x-axis between $-R$ and $+R$.

6
Applications of contour integration

6.1 Introduction

Using the basic theorems of the last chapter, in particular the Cauchy Residue Theorem (5.229), we may now apply contour integration to two specific problems: (i) the calculation of real integrals, and (ii) the summation of infinite series.

6.2 Calculation of real integrals

1. Integrals of the type $\int_0^{2\pi} f(\cos\theta, \sin\theta)\, d\theta$

The integrand $f(\cos\theta, \sin\theta)$ is first transformed into a function of $z = e^{i\theta}$ by writing

$$\cos\theta = \frac{e^{i\theta} + e^{-i\theta}}{2} = \frac{z + 1/z}{2}, \tag{6.1}$$

$$\sin\theta = \frac{e^{i\theta} - e^{-i\theta}}{2i} = \frac{z - 1/z}{2i}, \tag{6.2}$$

and $dz = ie^{i\theta}\, d\theta = iz\, d\theta$. The integral from $\theta = 0$ to $\theta = 2\pi$ is then equivalent to integrating around the unit circle C, defined by $z = e^{i\theta}$ or $|z| = 1$, and may be evaluated using the Cauchy Residue Theorem.

Example 1 Evaluate

$$I = \int_0^{2\pi} \frac{d\theta}{2 - \cos\theta}. \tag{6.3}$$

Choosing the unit circle C (as above) for which $z = e^{i\theta}$, we have

$$d\theta = dz/iz. \tag{6.4}$$

Using (6.1) and (6.4), (6.3) becomes

$$I = \oint_C \frac{dz/iz}{2 - \frac{1}{2}(z + 1/z)} = -\frac{2}{i}\oint_C \frac{dz}{z^2 - 4z + 1}. \tag{6.5}$$

Now the poles of the integrand occur where $z^2 - 4z + 1 = 0$, that is, at $z = 2 \pm \sqrt{3}$. Both are simple poles since $z^2 - 4z + 1 = [z - (2 - \sqrt{3})][z - (2 + \sqrt{3})]$, but only the pole at $z = 2 - \sqrt{3}$ is within C (see Figure 6.1). Hence, to evaluate the integral by the Cauchy Residue Theorem, we require only the residue at $z = 2 - \sqrt{3}$ which is (by (5.183))

$$\lim_{z \to 2-\sqrt{3}} \left\{[z - (2 - \sqrt{3})]\frac{1}{[z - (2 - \sqrt{3})][z - (2 + \sqrt{3})]}\right\}$$

$$= \frac{1}{(2 - \sqrt{3}) - (2 + \sqrt{3})} = -\frac{1}{2\sqrt{3}}. \tag{6.6}$$

Hence

$$\oint_C \frac{dz}{z^2 - 4z + 1} = 2\pi i\left(-\frac{1}{2\sqrt{3}}\right), \tag{6.7}$$

by the Cauchy Residue Theorem (5.229). Finally

$$\int_0^{2\pi} \frac{d\theta}{2 - \cos\theta} = -\frac{2}{i}2\pi i\left(-\frac{1}{2\sqrt{3}}\right) = \frac{2\pi}{\sqrt{3}}. \tag{6.8}$$

Example 2 Evaluate

$$I = \int_0^{2\pi} \frac{\cos(2\theta)}{5 - 4\cos\theta}\, d\theta. \tag{6.9}$$

Figure 6.1

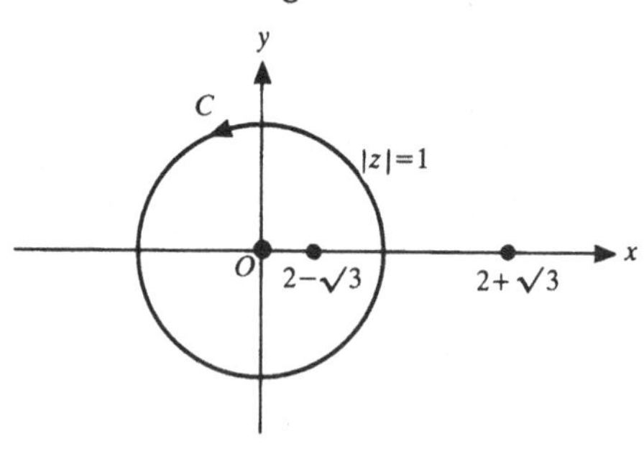

Using the fact that

$$\cos(2\theta) = \frac{e^{2i\theta} + e^{-2i\theta}}{2} = \frac{z^2 + 1/z^2}{2}, \tag{6.10}$$

together with (6.1) and (6.4), I becomes

$$\oint_C \frac{z^2 + 1/z^2}{2\left[5 - 4\left(\frac{z + 1/z}{2}\right)\right]} \frac{dz}{iz} = -\frac{1}{4i} \oint_C \frac{z^4 + 1}{z^2(z - \frac{1}{2})(z - 2)} dz, \tag{6.11}$$

where the contour C is the circle $|z| = 1$. The poles of the integrand are as follows:

(i) a pole of order 2 at $z = 0$;
(ii) a simple pole at $z = \frac{1}{2}$;
(iii) a simple pole at $z = 2$.

Of these, only the first two lie within C (see Figure 6.2) and hence we need the residues at these two poles only.

Residue at $z = 0$ Expanding the integrand as a Laurent series about $z = 0$, we have

$$\frac{(z^4 + 1)}{z^2}(z - \tfrac{1}{2})^{-1}(z - 2)^{-1} = \frac{(z^4 + 1)}{z^2}(1 - 2z)^{-1}(1 - z/2)^{-1}$$

$$= \frac{(z^4 + 1)}{z^2}(1 + 2z + 4z^2 + \ldots)(1 + z/2 + z^2/4 + \ldots). \tag{6.12}$$

The coefficient of $1/z$ is therefore $2 + \frac{1}{2} = \frac{5}{2}$, which is the residue at $z = 0$. Alternatively, since $z = 0$ is a pole of order 2, we could have used (5.206) with $m = 2$ to obtain the same result.

Figure 6.2

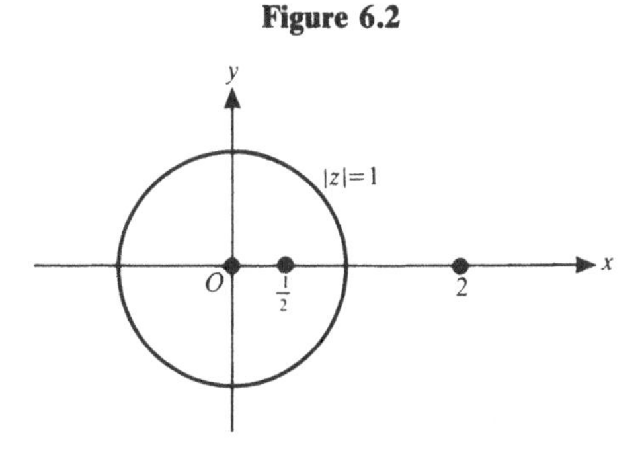

Residue at $z=\frac{1}{2}$ Here the pole is simple and the residue is (by (5.183))

$$\lim_{z\to\frac{1}{2}}\left[(z-\tfrac{1}{2})\frac{(z^4+1)}{z^2(z-\frac{1}{2})(z-2)}\right]=\frac{\frac{1}{16}+1}{\frac{1}{4}(-\frac{3}{2})}=-\tfrac{17}{6}. \tag{6.13}$$

Finally then, summing the two residues within C,

$$\oint_C \frac{z^4+1}{z^2(z-\frac{1}{2})(z-2)}\,dz = 2\pi i(\tfrac{5}{2}-\tfrac{17}{6}) = -\frac{2\pi i}{3}, \tag{6.14}$$

and consequently, by (6.11),

$$\int_0^{2\pi}\frac{\cos(2\theta)}{5-4\cos\theta}\,d\theta = -\frac{1}{4i}\left(-\frac{2\pi i}{3}\right)=\frac{\pi}{6}. \quad ◢ \tag{6.15}$$

2. *Integrals of the type* $\int_{-\infty}^{\infty} f(x)\,dx$

We replace x by z and consider the closed contour C in the upper half-plane consisting of the semicircle C' of radius R, for which $z=Re^{i\theta}$, $0\leqslant\theta\leqslant\pi$, and the part of the x-axis from $-R$ to $+R$ (see Figure 6.3). We further assume that the single-valued function $f(z)$ satisfies two conditions:

(i) $f(z)$ is analytic in the upper half-plane except for a finite number of poles, none of which lie on the real axis (the case of poles on the real axis is dealt with in Section 6.4);

(ii) $|f(z)|\leqslant M/R^2$, where $M>0$ is some constant, as $|z|=R$ tends to infinity on C' (the curved part of C).

Now, by the Cauchy Residue Theorem,

$$\oint_C f(z)\,dz = \int_{-R}^{R} f(x)\,dx + \int_{C'} f(z)\,dz$$
$$= 2\pi i \times (\text{sum of residues at the poles of } f(z) \text{ inside } C). \tag{6.16}$$

Figure 6.3

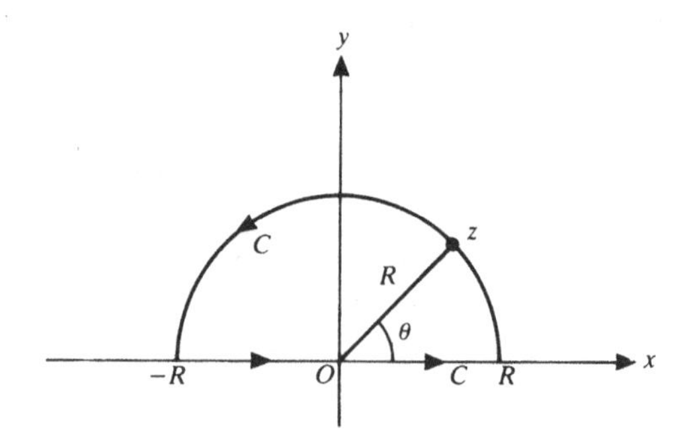

Further

$$\int_{-\infty}^{\infty} f(x)\, dx = \lim_{R\to\infty} \int_{-R}^{R} f(x)\, dx. \tag{6.17}$$

Hence, provided

$$\lim_{R\to\infty} \int_{C'} f(z)\, dz = 0, \tag{6.18}$$

as we expand the contour by letting $R \to \infty$, (6.16) gives (since, as $R \to \infty$, the interior of C becomes the upper half-plane)

$$\int_{-\infty}^{\infty} f(x)\, dx = 2\pi \mathrm{i} \times (\text{sum of residues at the poles of } f(z) \text{ in the upper half-plane}). \tag{6.19}$$

Condition (ii) above is sufficient to ensure that (6.18) is satisfied, for

$$\begin{aligned} \lim_{R\to\infty} \left| \int_{C'} f(z)\, dz \right| &= \lim_{R\to\infty} \left| \int_0^{\pi} f(R\mathrm{e}^{\mathrm{i}\theta}) \mathrm{i} R \mathrm{e}^{\mathrm{i}\theta}\, d\theta \right| \\ &\leqslant \lim_{R\to\infty} \int_0^{\pi} |f(R\mathrm{e}^{\mathrm{i}\theta})|\, R\, d\theta \\ &\leqslant \lim_{R\to\infty} \int_0^{\pi} \frac{M}{R^2} R\, d\theta = 0, \end{aligned} \tag{6.20}$$

where we have used the fact that the modulus of the integral of a function is less than or equal to the integral of the modulus of that function. The vanishing of the modulus of the integral on C' as $R \to \infty$ implies that the integral itself is zero.

Example 3 To evaluate

$$\int_{-\infty}^{\infty} \frac{dx}{x^4 + a^4}, \tag{6.21}$$

where a is a real positive constant.

Consider

$$\oint_C \frac{dz}{z^4 + a^4}, \tag{6.22}$$

where C is the closed contour of Figure 6.3. The conditions (i) and (ii) above are satisfied since $f(z) = (z^4 + a^4)^{-1}$ has simple poles where $z^4 = -a^4$, and $|f(z)|$ behaves like R^{-4} for large $|z| = R$. The four poles are at

$$z_k = a\mathrm{e}^{\mathrm{i}(\pi+2k\pi)/4}, \quad k = 0, 1, 2, 3, \tag{6.23}$$

and only two of these lie in the upper half-plane:

$$z_0 = a\mathrm{e}^{\mathrm{i}\pi/4}, \quad z_1 = a\mathrm{e}^{3\pi\mathrm{i}/4}. \tag{6.24}$$

To find the residues at these poles, we use (5.200) with $p(z) = 1$ and $q(z) = z^4 + a^4$. Hence the residue at $z_0 = a\mathrm{e}^{\mathrm{i}\pi/4}$ is

$$\left.\frac{p(z)}{q'(z)}\right|_{z_0} = \left.\frac{1}{4z^3}\right|_{z_0} = \frac{1}{4a^3}\,\mathrm{e}^{-3\pi\mathrm{i}/4}, \tag{6.25}$$

whilst that at $z_1 = a\mathrm{e}^{3\pi\mathrm{i}/4}$ is

$$\left.\frac{p(z)}{q'(z)}\right|_{z_1} = \left.\frac{1}{4z^3}\right|_{z_1} = \frac{1}{4a^3}\,\mathrm{e}^{-9\pi\mathrm{i}/4}. \tag{6.26}$$

Finally, using (6.19), we find

$$\begin{aligned}\int_{-\infty}^{\infty} \frac{dx}{x^4 + a^4} &= 2\pi\mathrm{i}\left(\frac{1}{4a^3}\,\mathrm{e}^{-3\pi\mathrm{i}/4} + \frac{1}{4a^3}\,\mathrm{e}^{-9\pi\mathrm{i}/4}\right) \\ &= \frac{2\pi\mathrm{i}}{4a^3}\left[\frac{(-1-\mathrm{i})}{\sqrt{2}} + \frac{(1-\mathrm{i})}{\sqrt{2}}\right] \\ &= \frac{\pi}{\sqrt{2}\,a^3}. \quad ◢\end{aligned} \tag{6.27}$$

3. *Integrals of the type $\int_{-\infty}^{\infty} f(x)\mathrm{e}^{\mathrm{i}kx}\,dx$ $(k > 0)$*

These integrals occur in the theory of Fourier transforms (see Chapter 7), and can be evaluated using the same contour C as in Figure 6.3. We assume that

(i) $f(z)$ is analytic in the upper half-plane except for a finite number of poles, none of which lie on the real axis (as before);

(ii) $|f(z)| \leq M/R$, where $M > 0$ is some constant, as $|z| = R$ tends to infinity on C'.

Condition (ii) ensures that the integral of $f(z)\mathrm{e}^{\mathrm{i}kz}$ on C' tends to zero as $R \to \infty$, for

$$\begin{aligned}&\lim_{R\to\infty}\left|\int_{C'} f(z)\mathrm{e}^{\mathrm{i}kz}\,dz\right| \\ &\leq \lim_{R\to\infty}\int_0^{\pi} |f(R\mathrm{e}^{\mathrm{i}\theta})|\,|\mathrm{e}^{\mathrm{i}kR\mathrm{e}^{\mathrm{i}\theta}}|\,R\,d\theta \\ &\leq \lim_{R\to\infty}\int_0^{\pi} \frac{M}{R}\,\mathrm{e}^{-kR\sin\theta}R\,d\theta = 2M\lim_{R\to\infty}\int_0^{\pi/2} \mathrm{e}^{-kR\sin\theta}\,d\theta.\end{aligned} \tag{6.28}$$

Since $\sin\theta \geqslant 2\theta/\pi$ for $0 \leqslant \theta \leqslant \pi/2$, the last integral in (6.28) is less than or equal to

$$2M \lim_{R\to\infty} \int_0^{\pi/2} e^{-2kR\theta/\pi}\, d\theta = \lim_{R\to\infty} \frac{M\pi}{kR}(1 - e^{-kR}), \tag{6.29}$$

which tends to zero as $R \to \infty$. The vanishing of the modulus of the integral on C' as $R \to \infty$ implies that the integral itself is zero. Using the Cauchy Residue Theorem for the contour C, as in ***2*** of this section, we have finally

$$\int_{-\infty}^{\infty} f(x)e^{ikx}\, dx = 2\pi i \times (\text{sum of residues of } f(z)e^{ikz} \text{ at the poles of } f(z) \text{ in the upper half-plane}). \tag{6.30}$$

Example 4 Evaluate

$$I = \int_0^{\infty} \frac{\cos(kx)}{x^2 + a^2}\, dx, \tag{6.31}$$

where $k > 0$, $a > 0$.

Here

$$I = \tfrac{1}{2} \operatorname{Re} \int_{-\infty}^{\infty} \frac{e^{ikx}}{x^2 + a^2}\, dx, \tag{6.32}$$

where Re stands for the real part of the integral. The integral in (6.32) now has the form of the left-hand side of (6.30). Consequently we consider

$$\oint_C \frac{e^{ikz}}{z^2 + a^2}\, dz, \tag{6.33}$$

where C is the contour of Figure 6.3, as before. The conditions (i) and (ii) are satisfied since $|f(z)| = |1/(z^2 + a^2)|$ behaves like R^{-2} on $|z| = R$ as $R \to \infty$. The poles of $f(z)$ are at $z = \pm ia$ of which only $z = ia$ lies in the upper half-plane. The residue at this pole is (using (5.183))

$$\lim_{z\to ia} \left[(z - ia) \frac{e^{ikz}}{(z - ia)(z + ia)} \right] = \frac{e^{-ka}}{2ia}. \tag{6.34}$$

Hence

$$\int_{-\infty}^{\infty} \frac{e^{ikx}}{x^2 + a^2}\, dx = 2\pi i \left(\frac{e^{-ka}}{2ia} \right) = \frac{\pi}{a} e^{-ka}, \tag{6.35}$$

and, since this is real, (6.32) gives

$$I = \frac{\pi}{2a} e^{-ka}. \tag{6.36}$$

We note here that if $k<0$ then the integral of (6.33) becomes unbounded on the curved part C' since

$$|e^{ikRe^{i\theta}}| = e^{-kR\sin\theta} \tag{6.37}$$

which, for $k<0$ and $0<\theta<\pi$, diverges as $R\to\infty$. In this case we use the contour in the lower half-plane (see Figure 6.4), the integral on the semicircular part tending to zero as $R\to\infty$ by an analysis similar to that in (6.28) and (6.29). Calculating the residue at $z=-ia$, we find

$$I=\frac{\pi}{2a}e^{ka}, \quad k<0. \tag{6.38}$$

Hence, for all k,

$$\int_0^\infty \frac{\cos(kx)}{x^2+a^2}\,dx = \frac{\pi}{2a}e^{-|k|a}. \tag{6.39}$$

6.3 An alternative contour

The contours of Sections 6.1 and 6.2 have been either circles or semicircles but other contours may be more suitable for a given integral. The following example shows how a rectangular contour may be used to obtain the value of an integral from a known result, without expanding the contour to infinity in *all* directions.

Example 5 Suppose we want to evaluate

$$I=\int_{-\infty}^{\infty} e^{-x^2}\cos(2ax)\,dx, \tag{6.40}$$

where $a>0$ is a real constant. Attempting the method of the last section and replacing x by z, we find that on a semicircle $z=Re^{i\theta}$,

Figure 6.4

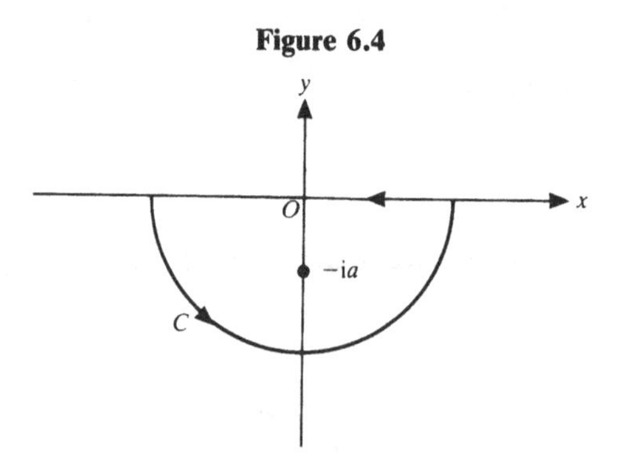

$|e^{-z^2}\cos(2az)| \to \infty$ as $|z| = R \to \infty$ when $\pi/4 < \theta < 3\pi/4$ (since $|e^{-z^2}| = e^{-R^2\cos(2\theta)}$ and $\cos(2\theta) < 0$ for $\pi/4 < \theta < 3\pi/4$). Hence the integral on this semicircle cannot tend to zero. Suppose instead we choose a rectangular contour C as shown in Figure 6.5, where the various straight portions are denoted by C_1, C_2, C_3 and C_4. Now consider the integral

$$\oint_C e^{-z^2}\,dz. \tag{6.41}$$

The function e^{-z^2} has no poles and it follows therefore from Cauchy's Theorem that this integral is zero. Hence

$$\oint_C e^{-z^2}\,dz = \int_{-x_1}^{x_1} e^{-x^2}\,dx + \int_0^{y_1} e^{-(x_1+iy)^2}\,i\,dy + \int_{x_1}^{-x_1} e^{-(x+iy_1)^2}\,dx + \int_{y_1}^{0} e^{-(-x_1+iy)^2}\,i\,dy = 0, \tag{6.42}$$

since on C_2, $e^{-z^2} = e^{-(x_1+iy)^2}$; on C_3, $e^{-z^2} = e^{-(x+iy_1)^2}$ and on C_4, $e^{-z^2} = e^{-(-x_1+iy)^2}$. We find, after some algebra, that

$$\int_{-x_1}^{x_1} e^{-x^2}\,dx - 2e^{-x_1^2}\int_0^{y_1} e^{y^2}\sin(2x_1y)\,dy - e^{y_1^2}\int_{-x_1}^{x_1} e^{-(x^2+2ixy_1)}\,dx = 0. \tag{6.43}$$

By putting $y_1 = a$ and letting $x_1 \to \infty$, we have

$$\int_{-\infty}^{\infty} e^{-x^2}\,dx = e^{a^2}\int_{-\infty}^{\infty} e^{-x^2}[\cos(2ax) - i\sin(2ax)]\,dx$$
$$= e^{a^2}\int_{-\infty}^{\infty} e^{-x^2}\cos(2ax)\,dx, \tag{6.44}$$

Figure 6.5

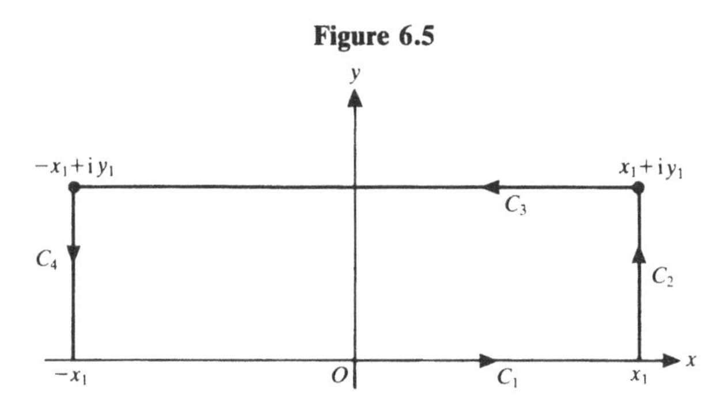

since $\sin(2ax)$ is an odd function. But, from (2.10), we deduce

$$\int_{-\infty}^{\infty} e^{-x^2}\,dx = \sqrt{\pi} \tag{6.45}$$

and hence

$$I = \int_{-\infty}^{\infty} e^{-x^2}\cos(2ax)\,dx = \sqrt{\pi}\,e^{-a^2}. \tag{6.46}$$

By differentiating with respect to a, we note that

$$\int_{-\infty}^{\infty} xe^{-x^2}\sin(2ax)\,dx = \sqrt{\pi}\,ae^{-a^2}. \quad ◢ \tag{6.47}$$

6.4 Poles on the real axis: the principal value of integrals

We have assumed in the previous sections that none of the poles of the integrand lie on the real axis. When a pole does lie on the real axis, we can still obtain a value for the integral $\int_{-\infty}^{\infty} f(x)\,dx$. However, since we cannot integrate through a singularity ($f(x)$ becomes infinite at a pole), we must assign a so-called principal value to the integral. Suppose a pole exists at $x = x_0$ ($a < x_0 < b$). Then the Cauchy principal value of $\int_a^b f(x)\,dx$ (denoted by **P**) is defined by

$$\mathbf{P}\int_a^b f(x)\,dx = \lim_{\delta\to 0}\left[\int_a^{x_0-\delta} f(x)\,dx + \int_{x_0+\delta}^{b} f(x)\,dx\right]. \tag{6.48}$$

Example 6 The integrand of

$$\int_0^{\infty} \frac{dx}{x^2-4} \tag{6.49}$$

is singular within the range of integration at $x = 2$. The principal value is therefore

$$\mathbf{P}\int_0^{\infty}\frac{dx}{x^2-4} = \lim_{\delta\to 0}\left(\int_0^{2-\delta}\frac{dx}{x^2-4} + \int_{2+\delta}^{\infty}\frac{dx}{x^2-4}\right) \tag{6.50}$$

$$= \lim_{\delta\to 0}\frac{1}{4}\left\{\left[\ln\left(\frac{2-x}{2+x}\right)\right]_0^{2-\delta} + \left[\ln\left(\frac{x-2}{x+2}\right)\right]_{2+\delta}^{\infty}\right\} \tag{6.51}$$

$$= \lim_{\delta\to 0}\frac{1}{4}\left\{\ln\left(\frac{\delta}{4-\delta}\right) + \ln\left(\frac{4+\delta}{\delta}\right)\right\} \tag{6.52}$$

$$= \lim_{\delta\to 0}\frac{1}{4}\ln\left(\frac{4+\delta}{4-\delta}\right) = 0. \quad ◢ \tag{6.53}$$

We now suppose that $\int_{-\infty}^{\infty} f(x)\,dx$ is to be evaluated using the techniques in **2** of Section 6.2. The function is assumed to have the same properties as stated in (i) and (ii) of that section, with the exception that a simple pole is now allowed at $x = x_0$ on the real axis. We indent the contour C by constructing a small semicircle γ of radius δ centred at x_0 (see Figure 6.6). Now (with C' being the curved part of C)

$$\oint_C f(z)\,dz = \int_{-R}^{x_0-\delta} f(x)\,dx + \int_{x_0+\delta}^{R} f(x)\,dx + \int_{C'} f(z)\,dz + \int_{\gamma} f(z)\,dz \tag{6.54}$$

$$= 2\pi\text{i} \times (\text{sum of residues at the poles of } f(z) \text{ inside } C). \tag{6.55}$$

To evaluate the last integral in (6.54), we use the Laurent expansion of $f(z)$ about $z = x_0$ in the form

$$f(z) = \frac{a_{-1}}{z - x_0} + a_0 + a_1(z - x_0) + \ldots. \tag{6.56}$$

Writing $z - x_0 = \delta e^{i\theta}$, we find

$$\int_{\gamma} f(z)\,dz = \int_{\pi}^{0} \frac{a_{-1}}{\delta e^{i\theta}} \delta i e^{i\theta}\,d\theta + a_0 \int_{\pi}^{0} \delta i e^{i\theta}\,d\theta$$

$$+ \text{ terms involving higher powers of } \delta. \tag{6.57}$$

Consequently,

$$\lim_{\delta\to 0} \int_{\gamma} f(z)\,dz = -\pi i a_{-1} = -2\pi i\left(\frac{a_{-1}}{2}\right), \tag{6.58}$$

where a_{-1} is the residue at $z = x_0$. The basic results (6.54) and (6.55) become, as $R \to \infty$ and $\delta \to 0$, using (6.58),

$$\lim_{\substack{R\to\infty \\ \delta\to 0}} \left[\int_{-R}^{x_0-\delta} f(x)\,dx + \int_{x_0+\delta}^{R} f(x)\,dx\right] = 2\pi i \times (\text{sum of the residues at the}$$

Figure 6.6

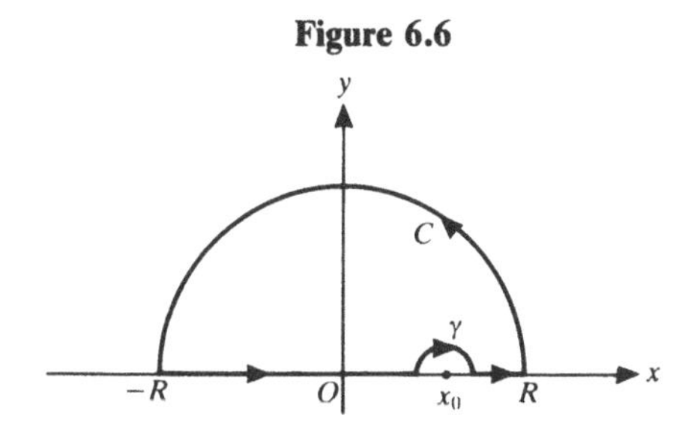

poles of $f(z)$ in the upper half-plane) $+ 2\pi i\left(\frac{a_{-1}}{2}\right)$, (6.59)

since $\int_{C'} f(z)\,dz \to 0$ as $R \to \infty$ (as before). Hence, using (6.48), the principal value is given by

$$\mathbf{P}\int_{-\infty}^{\infty} f(x)\,dx = 2\pi i \times (\text{sum of the residues at the poles of } f(z)$$

$$\text{in the upper half-plane}) + 2\pi i\left(\frac{a_{-1}}{2}\right). \quad (6.60)$$

If more than one pole exists on the real axis, (6.60) generalises to

$$\mathbf{P}\int_{-\infty}^{\infty} f(x)\,dx = 2\pi i$$

$$\times[(\text{sum of the residues at the poles of } f(z) \text{ in the upper half-plane})$$
$$+\tfrac{1}{2}(\text{sum of the residues at the poles of } f(z) \text{ on the real axis})]. \quad (6.61)$$

Example 7 A standard example is to evaluate

$$\int_{-\infty}^{\infty} \frac{e^{imx}}{x}\,dx, \quad (6.62)$$

where $m > 0$ is a real constant.

Writing the integral in complex form we have

$$\oint_C \frac{e^{imz}}{z}\,dz, \quad (6.63)$$

where C is the indented closed contour shown in Figure 6.7. The integrand has a pole at $z = 0$ and this is outside C because of the indentation (a small semicircle, radius δ, centred on $z = 0$). The function e^{imz}/z has no poles within C and the residue at $z = 0$ is

$$\lim_{z\to 0}\left(z e^{imz}/z\right) = 1. \quad (6.64)$$

Figure 6.7

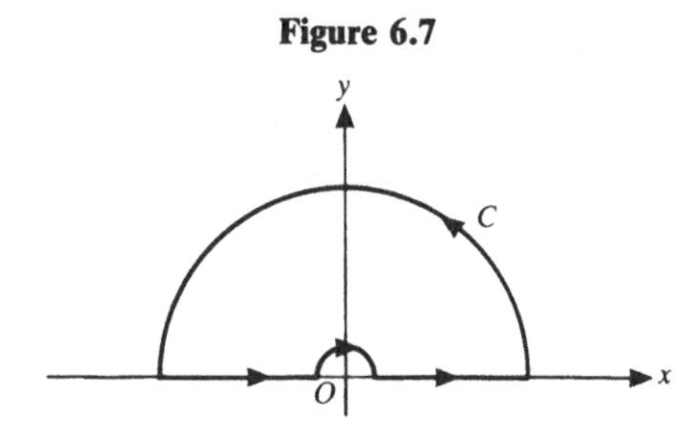

Hence, using (6.61), we have, on letting $R\to\infty$ and $\delta\to 0$,

$$\mathbf{P}\int_{-\infty}^{\infty}\frac{e^{imx}}{x}\,dx = 2\pi i(0+\tfrac{1}{2}\,.\,1) = \pi i. \tag{6.65}$$

Taking the real and imaginary parts of (6.65) we find

$$\mathbf{P}\int_{-\infty}^{\infty}\frac{\cos(mx)}{x}\,dx = 0 \tag{6.66}$$

and

$$\int_{-\infty}^{\infty}\frac{\sin(mx)}{x}\,dx = \pi. \tag{6.67}$$

We note that we have omitted the principal value symbol in (6.67) since $\sin(mx)/x$ does not diverge as $|x|\to 0$ and therefore does not have a pole on the real axis. If $m<0$, the contribution on the large semicircle does not tend to zero in the upper half-plane but, as before, we may use the contour in the lower half-plane shown in Figure 6.8. We find in the same way, by letting $R\to\infty$ and $\delta\to 0$, that

$$\int_{-\infty}^{\infty}\frac{\sin(mx)}{x}\,dx = -\pi, \quad m<0. \tag{6.68}$$

This integral will occur again in Chapter 7 and may be written as an integral from 0 to ∞,

$$\int_{0}^{\infty}\frac{\sin(mx)}{x}\,dx = \begin{cases}\pi/2, & m>0\\ 0, & m=0,\\ -\pi/2, & m<0\end{cases} \tag{6.69}$$

since the integrand in (6.68) is an even function of x. ◢

Example 8 By integrating $e^{imz}/z(z^2+a^2)$ around a semicircular contour in the upper half-plane, indented at the origin (see Figure

Figure 6.8

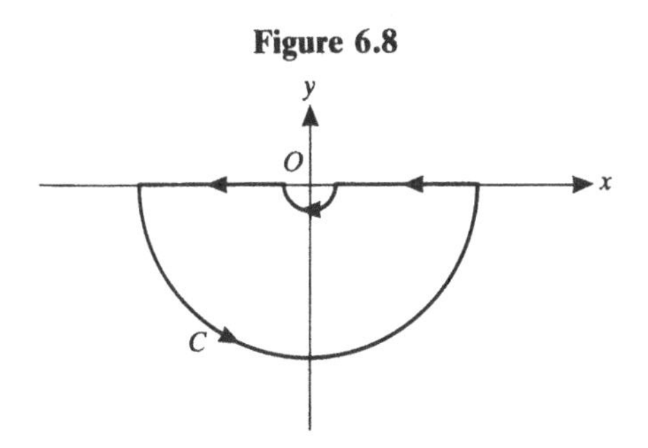

6.7), show using (6.61) that

$$\int_0^\infty \frac{\sin(mx)}{x(x^2+a^2)}\,dx = \frac{\pi}{2a^2}(1-\mathrm{e}^{-ma}), \tag{6.70}$$

where a and m are positive constants.

Since the integrand tends to zero sufficiently fast as $R \to \infty$,

$$\mathbf{P}\int_{-\infty}^{\infty} \frac{\mathrm{e}^{imx}}{x(x^2+a^2)}\,dx = 2\pi\mathrm{i} \times \text{(sum of the residues of the integrand at its poles in the upper half-plane)} + \pi\mathrm{i} \times \text{(residue of the integrand at } z=0). \tag{6.71}$$

In the upper half-plane there is a simple pole at $z = \mathrm{i}a$ with residue

$$\lim_{z \to \mathrm{i}a}\left[(z-\mathrm{i}a)\frac{\mathrm{e}^{imz}}{z(z+\mathrm{i}a)(z-\mathrm{i}a)}\right] = -\frac{1}{2a^2}\mathrm{e}^{-ma}. \tag{6.72}$$

For the simple pole at $z = 0$, the residue is

$$\lim_{z \to 0}\left[z\frac{\mathrm{e}^{imz}}{z(z^2+a^2)}\right] = \frac{1}{a^2}. \tag{6.73}$$

We have from (6.71) therefore

$$\mathbf{P}\int_{-\infty}^{\infty} \frac{\mathrm{e}^{imx}}{x(x^2+a^2)}\,dx = 2\pi\mathrm{i}\left(-\frac{1}{2a^2}\mathrm{e}^{-ma}\right) + \frac{\pi\mathrm{i}}{a^2}$$

$$= \frac{\pi\mathrm{i}}{a^2}(1-\mathrm{e}^{-ma}). \tag{6.74}$$

Taking the imaginary part of each side and noticing that the integrand is an even function, we have

$$\int_0^\infty \frac{\sin(mx)}{x(x^2+a^2)}\,dx = \frac{\pi}{2a^2}(1-\mathrm{e}^{-ma}). \tag{6.75}$$

By differentiating with respect to the parameter m we find

$$\int_0^\infty \frac{\cos(mx)}{(x^2+a^2)}\,dx = \frac{\pi}{2a}\mathrm{e}^{-ma}, \tag{6.76}$$

as in (6.36). ◢

6.5 Branch points and integrals of many-valued functions

Consider now, for example,

$$\int_0^\infty x^\alpha f(x)\,dx, \tag{6.77}$$

where α is a constant but not an integer. Writing

$$\oint_C z^{\alpha} f(z)\,dz, \tag{6.78}$$

we notice that z^{α} is a multi-valued function and that $z=0$ is a branch point. In Section 5.1, we discussed the functions $z^{\frac{1}{2}}$ and $z^{\frac{1}{3}}$ and found that if the argument of z is changed by 2π (in other words, if a branch point is encircled) then we obtain another branch of the function and a different value from the starting value. In using the contour integration technique, therefore, the contour must be chosen to exclude the branch point so avoiding multiple values. Some integrals involving multi-valued functions can be evaluated using a suitably indented semicircle, and we illustrate this with two examples.

Example 9 Consider

$$\int_0^{\infty} \frac{x^{\alpha}}{1+x^2}\,dx, \tag{6.79}$$

where $0<\alpha<1$.

Accordingly, we evaluate

$$\oint_C \frac{z^{\alpha}}{1+z^2}\,dz, \tag{6.80}$$

where C is the indented closed contour shown in Figure 6.7. The small circle, radius δ, centred on $z=0$ ensures that the branch point $z=0$ is outside the contour. As $R\to\infty$ on the large semicircle $z=Re^{i\theta}$, $\int |[z^{\alpha}/(1+z^2)]\,dz|$ behaves like $R^{\alpha-1}$ which tends to zero. On the small semicircle $z=\delta e^{i\theta}$

$$\int \frac{z^{\alpha}}{1+z^2}\,dz = \int \frac{i\delta^{\alpha+1}e^{i(1+\alpha)\theta}}{1+\delta^2 e^{2i\theta}}\,d\theta, \tag{6.81}$$

which tends to zero as $\delta\to 0$. Hence

$$\int_0^{\infty} \frac{x^{\alpha}}{1+x^2}\,dx + \int_{-\infty}^{0} \frac{x^{\alpha}}{1+x^2}\,dx = 2\pi i \times [\text{residue of } z^{\alpha}/(1+z^2) \text{ at } z=i]. \tag{6.82}$$

By changing x to $-x = xe^{\pi i}$ in the second integral and evaluating the residue of $z^{\alpha}/(1+z^2)$ at the simple pole $z=i$, we find

$$\int_0^{\infty} \frac{x^{\alpha}}{1+x^2}\,dx + \int_0^{\infty} \frac{(xe^{\pi i})^{\alpha}}{1+x^2}\,dx = 2\pi i\,\frac{(e^{\pi i/2})^{\alpha}}{2i}. \tag{6.83}$$

Finally then

$$(1+e^{\pi i\alpha})\int_0^\infty \frac{x^\alpha}{1+x^2}\,dx = \pi e^{\pi i\alpha/2}, \tag{6.84}$$

giving

$$\int_0^\infty \frac{x^\alpha}{1+x^2}\,dx = \frac{\pi e^{\pi i\alpha/2}}{1+e^{\pi i\alpha}} = \frac{\pi}{2\cos(\pi\alpha/2)}. \quad ◢ \tag{6.85}$$

Example 10 Consider

$$\int_0^\infty \frac{\ln x}{x^2+a^2}\,dx, \tag{6.86}$$

where $a>0$.

Using the same indented semicircular contour (see Figure 6.7) to avoid the branch point at $z=0$, we now examine

$$\oint_C \frac{\ln z}{z^2+a^2}\,dz. \tag{6.87}$$

On the large semicircle, $\int |[(\ln z)/(z^2+a^2)]\,dz|$ behaves like $R^{-1}\ln R$ which tends to zero as $R\to\infty$. On the small semicircle $z=\delta e^{i\theta}$,

$$\int \frac{\ln z}{z^2+a^2}\,dz = \int \frac{(\ln\delta + i\theta)i\delta e^{i\theta}}{\delta^2 e^{2i\theta}+a^2}\,d\theta \tag{6.88}$$

which tends to zero as $\delta\to 0$. Hence

$$\int_0^\infty \frac{\ln x}{x^2+a^2}\,dx + \int_{-\infty}^0 \frac{\ln x}{x^2+a^2}\,dx = 2\pi i \times [\text{residue of } \ln z/(z^2+a^2) \text{ at } z=ia], \tag{6.89}$$

so that changing x to $-x = xe^{\pi i}$ in the second integral, we obtain

$$\int_0^\infty \frac{\ln x}{x^2+a^2}\,dx + \int_0^\infty \frac{\ln(xe^{\pi i})}{x^2+a^2}\,dx = 2\pi i\,\frac{\ln(ia)}{2ia}. \tag{6.90}$$

Expanding the logarithms into real and imaginary parts, and taking the principal value of $\ln(ia)$ since all arguments within C lie between 0 and π, we have

$$2\int_0^\infty \frac{\ln x}{x^2+a^2}\,dx + \pi i\int_0^\infty \frac{dx}{x^2+a^2} = \frac{\pi}{a}\left(\ln a + i\frac{\pi}{2}\right). \tag{6.91}$$

The real part of (6.91) gives

$$\int_0^\infty \frac{\ln x}{x^2+a^2}\,dx = \frac{\pi}{2a}\ln a. \quad ◢ \tag{6.92}$$

It should be noted in the above two examples that the integral from $-\infty$ to 0 included, or was a multiple of, the required integral from 0 to ∞. This arose because the single-valued parts of the integrands ($1/(1+x^2)$ and $1/(x^2+a^2)$, respectively) were even functions of x. If this is not the case then we require an alternative contour which also excludes the branch point. A suitable contour is shown in Figure 6.9. The contour C is made up of the large broken circle (radius R), the two straight lines along the x-axis separated by an infinitesimal distance (the 'branch cut'), and the small broken circle (radius δ) surrounding the branch point $z=0$. We shall see that, due to the multi-valued nature of the integrand, the contributions above and below the branch cut do not cancel. The following example illustrates the use of this contour.

Example 11 Evaluate

$$I = \int_0^\infty \frac{\sqrt{x}}{1+x^3}\,dx. \tag{6.93}$$

We therefore evaluate

$$\oint_C \frac{\sqrt{z}}{1+z^3}\,dz, \tag{6.94}$$

where C is the contour of Figure 6.9. Both contributions from the large circle of radius R and the small circle of radius δ tend to zero (as $R\to\infty$ and $\delta\to 0$, respectively). The contributions from above and below the cut are, respectively, the required integral I in (6.93) and (putting $z = xe^{2\pi i}$ on the line below the cut)

$$\int_\infty^0 \frac{\sqrt{(xe^{2\pi i})}}{1+(xe^{2\pi i})^3}\,e^{2\pi i}\,dx = -e^{\pi i}\int_0^\infty \frac{\sqrt{x}}{1+x^3}\,dx = I. \tag{6.95}$$

Figure 6.9

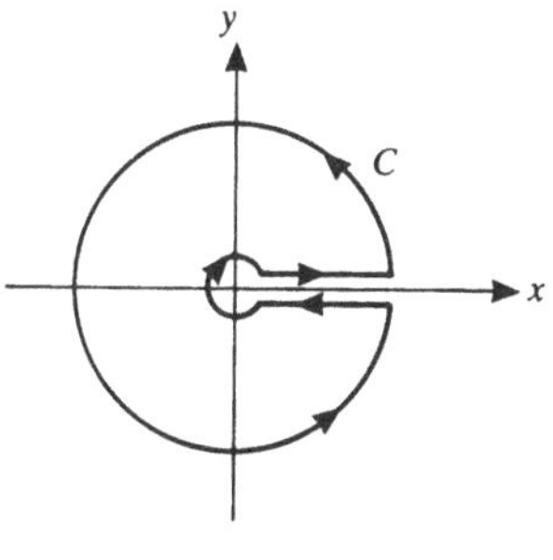

y

C

x

Hence

$$\oint_C \frac{\sqrt{z}}{1+z^3}\,dz = 2I = 2\pi\mathrm{i}$$

$$\times\,[\text{sum of the residues of } \sqrt{z}/(1+z^3) \text{ at all its poles}]. \quad (6.96)$$

The poles occur at $z^3 = -1 = \mathrm{e}^{\pi\mathrm{i}+2k\pi\mathrm{i}}$, giving

$$\begin{aligned} z_1 &= \mathrm{e}^{\pi\mathrm{i}/3}, \\ z_2 &= \mathrm{e}^{\pi\mathrm{i}} = -1, \\ z_3 &= \mathrm{e}^{5\pi\mathrm{i}/3}. \end{aligned} \quad (6.97)$$

Since these are simple poles, the residues are, by (5.200), $(\sqrt{z}/3z^2)_{z=z_k} = \frac{1}{3}z_k^{-\frac{3}{2}}$, for $k = 1, 2, 3$. From (6.96) we have therefore

$$I = \frac{\pi\mathrm{i}}{3}(\mathrm{e}^{-\pi\mathrm{i}/2} + \mathrm{e}^{-3\pi\mathrm{i}/2} + \mathrm{e}^{-5\pi\mathrm{i}/2}) \quad (6.98)$$

$$= \frac{\pi\mathrm{i}}{3}(-\mathrm{i}+\mathrm{i}-\mathrm{i}) = \frac{\pi}{3}. \quad ◢ \quad (6.99)$$

We note here that the integral in Example 9 can also be evaluated using the cut circular contour of Example 11. However, the logarithmic integral in Example 10 cannot be evaluated in this way. We leave the reader to attempt this and to verify that, although the contributions from above and below the branch cut do not entirely cancel, those terms containing the required integral do cancel.

6.6 Summation of series

Contour integration can be used to sum particular infinite series. Suppose $f(z)$ is a function which is analytic at the integers $z = 0, \pm1, \pm2, \ldots$ and tends to zero at least as fast as $|z|^{-2}$ as $|z| \to \infty$. Now consider the function $F(z) = \pi\cot(\pi z)f(z)$. This function $F(z)$ has simple poles at $z = n$ $(n = 0, \pm1, \pm2, \ldots)$ with residues (using (5.200) with $p(z) = \pi\cos(\pi z)f(z)$ and $q(z) = \sin(\pi z)$) given by

$$\left.\frac{\pi\cos(\pi z)f(z)}{(d/dz)\sin(\pi z)}\right|_{z=n} = f(n). \quad (6.100)$$

We now integrate $\pi\cot(\pi z)f(z)$ around a square S with corners at the points $z = (N+\frac{1}{2})(\pm1\pm\mathrm{i})$, where N is a positive integer (see Figure

6.10). Using the Cauchy Residue Theorem,

$$\oint_S \pi \cot(\pi z) f(z)\, dz = 2\pi \mathrm{i} \times [\text{sum of the residues at the poles of } \pi \cot(\pi z) f(z) \text{ inside the square}]$$

$$= 2\pi \mathrm{i}\left\{ \sum_{n=-N}^{+N} f(n) + \begin{bmatrix} \text{sum of the residues of } \pi \cot(\pi z) f(z) \\ \text{at the poles of } f(z) \text{ inside the square} \end{bmatrix} \right\}, \quad (6.101)$$

where the first term is the sum of the residues of $\pi \cot(\pi z) f(z)$ at the poles of $\cot(\pi z)$. On the horizontal sides of S, $z = x + iy$ where $y = N + \frac{1}{2} > \frac{1}{2}$. Then

$$|\cot(\pi z)| = \left| \frac{\mathrm{e}^{\mathrm{i}\pi z} + \mathrm{e}^{-\mathrm{i}\pi z}}{\mathrm{e}^{\mathrm{i}\pi z} - \mathrm{e}^{-\mathrm{i}\pi z}} \right| \leqslant \left| \frac{|\mathrm{e}^{\mathrm{i}\pi z}| + |\mathrm{e}^{-\mathrm{i}\pi z}|}{|\mathrm{e}^{\mathrm{i}\pi z}| - |\mathrm{e}^{-\mathrm{i}\pi z}|} \right|$$

$$= \left| \frac{\mathrm{e}^{-\pi y} + \mathrm{e}^{\pi y}}{\mathrm{e}^{-\pi y} - \mathrm{e}^{\pi y}} \right| = \coth(\pi\,|y|) < \coth(\pi/2), \quad (6.102)$$

since $y > \frac{1}{2}$. Further, on the vertical sides of S, $z = \pm(N + \frac{1}{2}) + \mathrm{i}y$ so that

$$|\cot(\pi z)| = |\cot\{\pi[\pm(N + \tfrac{1}{2}) + \mathrm{i}y]\}| = |\cot[\pi(\pm\tfrac{1}{2} + \mathrm{i}y)]|$$

$$= |\tan(\pi \mathrm{i}y)| = |\tanh(\pi y)| < \coth(\pi/2). \quad (6.103)$$

Hence, from (6.102) and (6.103), $\cot(\pi z)$ is bounded on S as $N \to \infty$ and consequently $\int |\pi \cot(\pi z) f(z)\, dz|$ tends to zero on S as $N \to \infty$ because of the behaviour of $f(z)$. Letting $N \to \infty$ in (6.101), we see that the integral on the left-hand side tends to zero giving

$$\sum_{n=-\infty}^{+\infty} f(n) = -[\text{sum of the residues of } \pi \cot(\pi z) f(z) \text{ at the poles of } f(z)], \quad (6.104)$$

Figure 6.10

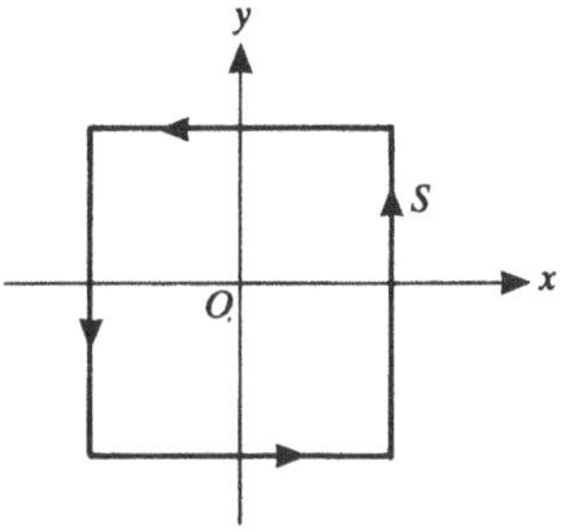

which enables the series on the left-hand side of (6.104) to be summed. Similarly, if we consider the function $\pi \operatorname{cosec}(\pi z) f(z)$, we find that the residues at the integers are $(-1)^n f(n)$. The corresponding result to (6.104) is

$$\sum_{n=-\infty}^{+\infty} (-1)^n f(n)$$
$$= -[\text{sum of the residues of } \pi \operatorname{cosec}(\pi z) f(z) \text{ at the poles of } f(z)], \tag{6.105}$$

which enables series with alternating signs to be summed.

Example 12 Consider

$$\sum_{n=0}^{\infty} \frac{1}{n^2+1} = 1 + \tfrac{1}{2} + \tfrac{1}{5} + \tfrac{1}{10} + \ldots. \tag{6.106}$$

To express this in the form of a sum from $-\infty$ to $+\infty$, we write

$$\sum_{n=0}^{\infty} \frac{1}{n^2+1} = 1 + \sum_{n=1}^{\infty} \frac{1}{n^2+1} = \tfrac{1}{2} + \tfrac{1}{2} \sum_{n=-\infty}^{+\infty} \frac{1}{n^2+1}. \tag{6.107}$$

Now, using (6.104), we have

$$\sum_{n=-\infty}^{+\infty} \frac{1}{n^2+1} = -[\text{sum of residues of } \pi \cot(\pi z)/(z^2+1) \text{ at } z = \pm \mathrm{i}]. \tag{6.108}$$

Both poles at $z = \pm\mathrm{i}$ are simple. The residue at $z = +\mathrm{i}$ is therefore

$$\frac{\pi \cot(\pi \mathrm{i})}{2\mathrm{i}} = -\frac{\pi}{2} \coth \pi, \tag{6.109}$$

whilst that at $z = -\mathrm{i}$ is

$$\frac{\pi \cot(-\pi \mathrm{i})}{-2\mathrm{i}} = -\frac{\pi}{2} \coth \pi. \tag{6.110}$$

Hence, from (6.107) and (6.108),

$$\sum_{n=0}^{\infty} \frac{1}{n^2+1} = \tfrac{1}{2} + \frac{\pi}{2} \coth \pi. \tag{6.111}$$

Similarly for the series

$$\sum_{n=0}^{\infty} \frac{(-1)^n}{n^2+1} = 1 - \tfrac{1}{2} + \tfrac{1}{5} - \tfrac{1}{10} + \ldots, \tag{6.112}$$

we write

$$\sum_{n=0}^{\infty} \frac{(-1)^n}{n^2+1} = \tfrac{1}{2} + \tfrac{1}{2} \sum_{n=-\infty}^{+\infty} \frac{(-1)^n}{n^2+1}. \tag{6.113}$$

Using (6.104) we have

$$\sum_{n=-\infty}^{+\infty} \frac{(-1)^n}{n^2+1} = -[\text{sum of the residues of } \pi \operatorname{cosec}(\pi z)/(z^2+1) \text{ at } z = \pm \mathrm{i}] \tag{6.114}$$

$$= -\left(\frac{\pi \operatorname{cosec}(\pi \mathrm{i})}{2\mathrm{i}} + \frac{\pi \operatorname{cosec}(-\pi \mathrm{i})}{-2\mathrm{i}} \right) = \frac{\pi}{\sinh \pi}. \tag{6.115}$$

Hence, from (6.113),

$$\sum_{n=0}^{\infty} \frac{(-1)^n}{n^2+1} = \tfrac{1}{2} + \frac{\pi}{2 \sinh \pi}. \quad ◢ \tag{6.116}$$

Often the series we wish to sum begins with the $n=1$ term, and the function $f(z)$ has a pole at $z=0$, corresponding to the $n=0$ term. In this case, since the $n=0$ term diverges in the sums on the left-hand sides of (6.104) and (6.105), we omit the term $n=0$ and the residue at $z=0$ is included on the right-hand sides. We illustrate this with an example.

Example 13 Consider

$$\sum_{n=1}^{\infty} \frac{1}{n^2} = 1 + \tfrac{1}{4} + \tfrac{1}{9} + \tfrac{1}{16} + \ldots. \tag{6.117}$$

Then $f(z) = 1/z^2$ has a pole at $z=0$. Hence

$$\sum_{\substack{n=-\infty \\ n \neq 0}}^{\infty} \frac{1}{n^2} = -\left[\text{residue of } \frac{\pi \cot(\pi z)}{z^2} \text{ at } z = 0 \right], \tag{6.118}$$

since $z=0$ is the only pole of $f(z)$. To find the residue here, we expand $\pi \cot(\pi z)/z^2$ as a Laurent series about $z=0$:

$$\frac{\pi \cot(\pi z)}{z^2} = \frac{\pi \cos(\pi z)}{z^2 \sin(\pi z)} = \frac{\pi}{z^2} \frac{\left(1 - \dfrac{\pi^2 z^2}{2!} + \ldots\right)}{\left(\pi z - \dfrac{\pi^3 z^3}{3!} + \ldots\right)}$$

$$= \frac{1}{z^3}\left(1 - \frac{\pi^2 z^2}{2} + \ldots\right)\left(1 - \frac{\pi^2 z^2}{6} + \ldots\right)^{-1}$$

$$= \frac{1}{z^3}\left(1 - \frac{\pi^2 z^2}{2} + \ldots\right)\left(1 + \frac{\pi^2 z^2}{6} + \ldots\right)$$

$$= \frac{1}{z^3}\left(1 - \frac{\pi^2 z^2}{3} + \ldots\right). \tag{6.119}$$

The residue (the coefficient of $1/z$) at this pole (of order 3) is therefore $-\pi^2/3$. From (6.118) we have

$$\sum_{\substack{n=-\infty \\ n\neq 0}}^{\infty} 1/n^2 = \pi^2/3, \tag{6.120}$$

so that

$$\sum_{n=1}^{\infty} 1/n^2 = \pi^2/6. \tag{6.121}$$

Similarly for the series

$$\sum_{n=1}^{\infty} (-1)^{n-1}/n^2 = 1 - \tfrac{1}{4} + \tfrac{1}{9} - \tfrac{1}{16} + \ldots, \tag{6.122}$$

we have

$$\sum_{\substack{n=-\infty \\ n\neq 0}}^{\infty} (-1)^n/n^2 = -[\text{residue of } \pi \operatorname{cosec}(\pi z)/z^2 \text{ at } z=0]. \tag{6.123}$$

Expanding about $z=0$ to find the residue, we have

$$\begin{aligned}\frac{\pi \operatorname{cosec}(\pi z)}{z^2} &= \frac{\pi}{z^2}\left(\pi z - \frac{\pi^3 z^3}{3!} + \ldots\right)^{-1} \\ &= \frac{1}{z^3}\left(1 - \frac{\pi^2 z^2}{6} + \ldots\right)^{-1} \\ &= \frac{1}{z^3}\left(1 + \frac{\pi^2 z^2}{6} + \ldots\right).\end{aligned} \tag{6.124}$$

Hence the residue (the coefficient of $1/z$) is $\pi^2/6$ and consequently

$$\sum_{n=1}^{\infty} \frac{(-1)^{n-1}}{n^2} = -\tfrac{1}{2} \sum_{\substack{n=-\infty \\ n\neq 0}}^{\infty} \frac{(-1)^n}{n^2} = \frac{\pi^2}{12}. \quad ◢ \tag{6.125}$$

Problems 6

1. Evaluate by contour integration around the unit circle $|z|=1$:

$$\text{(i)} \quad \int_0^{2\pi} \frac{\cos(3\theta)}{5-4\cos\theta}\, d\theta, \qquad \text{(ii)} \quad \int_0^{2\pi} \frac{\sin^2\theta}{2-\cos\theta}\, d\theta.$$

2. Show by contour integration that

$$\int_0^{2\pi} \frac{d\theta}{1-2a\cos\theta + a^2} = \frac{2\pi}{1-a^2}, \quad |a|<1.$$

Hence show that

$$\int_0^{2\pi} \frac{d\theta}{(1+a^2)^2 - 4a^2\cos^2\theta} = \frac{2\pi}{1-a^4}, \quad |a| < 1.$$

3. Evaluate, using a semicircular contour in the upper half-plane centred at the origin,

(i) $\displaystyle\int_{-\infty}^{\infty} \frac{dx}{1+x^6}$, (ii) $\displaystyle\int_{-\infty}^{\infty} \frac{dx}{x^2+2x+2}$,

(iii) $\displaystyle\int_0^{\infty} \frac{x^2\,dx}{(x^2+1)(x^2+4)}$, (iv) $\displaystyle\int_{-\infty}^{\infty} \frac{x^2\,dx}{(x^2+1)^2}$.

4. By integrating

$$\frac{e^{iz}}{(z^2+a^2)(z^2+b^2)},$$

where $a > b > 0$, around a semicircle in the upper half-plane centred at the origin, show that

$$\int_{-\infty}^{\infty} \frac{\cos x}{(x^2+a^2)(x^2+b^2)}\,dx = \frac{\pi}{(a^2-b^2)}\left(\frac{e^{-b}}{b} - \frac{e^{-a}}{a}\right).$$

5. By integrating the function

$$\frac{e^{3iz}}{(z^2+4)^2}$$

around a semicircular contour in the upper half-plane centred at the origin, show that

$$\int_0^{\infty} \frac{\cos(3x)}{(x^2+4)^2}\,dx = \frac{7\pi}{32}e^{-6}.$$

6. Find the principal value of the integral

$$\int_{-\infty}^{\infty} \frac{dx}{(x+1)(x^2+2)}.$$

7. Show that, if $a \neq b$, the function

$$\frac{e^{ibz} - e^{iaz}}{z^2}$$

has a simple pole on the real axis. By integrating this function around a suitably indented semicircular contour, show that

$$\int_{-\infty}^{\infty} \frac{\cos(bx) - \cos(ax)}{x^2}\,dx = \pi(a-b),$$

where a and b are positive constants.

8. Show that the zeros of $1-2e^{-iz}$ lie in the lower half-plane ($\operatorname{Im} z<0$). Evaluate

$$I=\oint_C \frac{z\,dz}{1-2e^{-iz}},$$

where C is the rectangular contour with corners at the points $(-\pi, 0)$, $(\pi, 0)$, (π, iK) and $(-\pi, iK)$. Show, by letting $K\to\infty$, that

$$\int_0^{\pi}\frac{x\sin x}{5-4\cos x}\,dx=\frac{\pi}{2}\ln(\tfrac{3}{2}).$$

9. Evaluate

$$\oint_C \frac{z^{\alpha-1}}{1+z}\,dz,$$

where $0<\alpha<1$, where C is the contour in Figure 6.9 consisting of concentric circles of radii $R>1$ and $\delta<1$ with centres at the origin joined by a cut along the positive real axis. Deduce, that, as $R\to\infty$ and $\delta\to 0$,

$$\int_0^{\infty}\frac{x^{\alpha-1}}{1+x}\,dx=\frac{\pi}{\sin(\pi\alpha)}.$$

10. Show, by using the same contour as in Problem 9, that

(i)
$$\int_0^{\infty}\frac{\sqrt{x}}{(1+x^2)^2(1+x)}\,dx=\frac{\pi}{2\sqrt{2}}\left(1-\frac{1}{\sqrt{2}}\right),$$

(ii)
$$\int_0^{\infty}\frac{\sqrt{x}\ln x}{(1+x)^2}\,dx=\pi.$$

11. Evaluate

(i) $\sum_{n=0}^{\infty} 1/(n^2+1)^2$, (ii) $\sum_{n=1}^{\infty}(-1)^n/n^4$.

7
Laplace and Fourier transforms

7.1 Introduction

In the mathematical analysis of many linear problems it is often useful to define an integral transform of a function $f(x)$. The general form of such a transform is given by

$$\bar{f}(s) = \int_a^b f(x)K(s, x)\, dx, \tag{7.1}$$

where $\bar{f}(s)$ is the integral transform of $f(x)$ with respect to the kernel $K(s, x)$ (a given function of two variables x and s), a and b being real constants. There are a number of important transforms (for example, Laplace, Fourier, Hankel and Mellin) obtained by choosing different forms for $K(s, x)$ and different values for a and b. The operation of taking the integral transform, as in (7.1), exhibits a linearity property. Suppose we let $I\{\ \}$ denote the operation of taking the transform of whatever function occurs inside the curly brackets. Then

$$I\{f(x)\} = \bar{f}(s) = \int_a^b f(x)K(s, x)\, dx \tag{7.2}$$

and clearly

$$I\{\alpha f(x)\} = \alpha I\{f(x)\}, \tag{7.3}$$

whilst

$$I\{\alpha f(x) + \beta g(x)\} = \int_a^b [\alpha f(x) + \beta g(x)]K(s, x)\, dx \tag{7.4}$$

$$= \alpha I\{f(x)\} + \beta I\{g(x)\}, \tag{7.5}$$

where α and β are arbitrary constants. Equations (7.3) and (7.5) show that $I\{\ \}$ is a linear operator. In order that $f(x)$ may be obtained if $\bar{f}(s)$ is given, we now introduce the inverse operator $I^{-1}\{\ \}$ which is such that if

$$I\{f(x)\} = \bar{f}(s) \tag{7.6}$$

then

$$f(x) = I^{-1}\{\bar{f}(s)\}. \tag{7.7}$$

Accordingly $II^{-1} = I^{-1}I = \mathbf{1}$ (the unit operator). It can be shown that I^{-1} is also a linear operator. In this chapter, we concentrate on two of the most commonly used transforms – the Laplace transform and the Fourier transform.

7.2 The Laplace transform

We suppose that $f(x)$ is defined for $x \geqslant 0$. The Laplace transform of $f(x)$ is then

$$\mathscr{L}\{f(x)\} = \bar{f}(s) = \int_0^\infty f(x)e^{-sx}\,dx. \tag{7.8}$$

In general, the variable s may be complex. We shall assume, however, that s is real until ***3*** of Section 7.8. It can be proved that provided $|f(x)e^{-\alpha x}| \leqslant M$ as $x \to \infty$, where M and α are suitable constants, then (7.8) will exist. Functions satisfying this condition are said to be of exponential order. We note that the function e^{x^2} is not of exponential order since $e^{x^2}e^{-\alpha x}$ increases without limit as $x \to \infty$ for all α. It is not possible, therefore, to define the Laplace transform of this function. The Laplace transforms of some elementary functions are given in Table 7.1. The conditions on s given in Table 7.1 ensure that the integral in (7.8) exists.

The proofs of these results all follow by integration from the basic definition (7.8), and from use of the linearity properties (7.3)–(7.5). For example

$$\mathscr{L}\{\sinh(ax)\} = \mathscr{L}\left\{\frac{e^{ax} - e^{-ax}}{2}\right\} = \tfrac{1}{2}\mathscr{L}\{e^{ax}\} - \tfrac{1}{2}\mathscr{L}\{e^{-ax}\}$$

$$= \tfrac{1}{2}\left(\frac{1}{s-a} - \frac{1}{s+a}\right) = \frac{a}{s^2 - a^2}, \tag{7.9}$$

using the standard transform of e^{ax} (see Table 7.1).

Table 7.1

$f(x)$	$\mathscr{L}\{f(x)\} = \bar{f}(s)$
1	$1/s \quad (s > 0)$
$x^n \ (n = 0, 1, 2, \ldots)$	$n!/s^{n+1} \quad (s > 0)$
e^{ax}	$\dfrac{1}{s-a} \quad (s > a)$
$\sin(ax)$	$\dfrac{a}{s^2 + a^2} \quad (s > 0)$
$\cos(ax)$	$\dfrac{s}{s^2 + a^2} \quad (s > 0)$
$\sinh(ax)$	$\dfrac{a}{s^2 - a^2} \quad (s > \lvert a \rvert)$
$\cosh(ax)$	$\dfrac{s}{s^2 - a^2} \quad (s > \lvert a \rvert)$

7.3 Three basic theorems

1. The Shift Theorem

This states that if

$$\mathscr{L}\{f(x)\} = \bar{f}(s) \tag{7.10}$$

then

$$\mathscr{L}\{e^{-ax}f(x)\} = \bar{f}(s + a). \tag{7.11}$$

Proof Using (7.8),

$$\mathscr{L}\{e^{-ax}f(x)\} = \int_0^\infty e^{-ax}e^{-sx}f(x)\,dx \tag{7.12}$$

$$= \int_0^\infty e^{-(s+a)x}f(x)\,dx = \bar{f}(s + a), \tag{7.13}$$

since this integral is again (7.8) with s replaced by $s + a$.

2.

If

$$\mathscr{L}\{f(x)\} = \bar{f}(s) \tag{7.14}$$

then

$$\mathscr{L}\{x^n f(x)\} = (-1)^n \frac{d^n}{ds^n}\bar{f}(s), \tag{7.15}$$

where $n = 1, 2, 3 \ldots.$

Proof Using (7.8),

$$\bar{f}(s) = \int_0^\infty e^{-sx} f(x)\, dx \tag{7.16}$$

so that

$$\frac{d^n}{ds^n}\bar{f}(s) = \frac{d^n}{ds^n}\int_0^\infty e^{-sx} f(x)\, dx \tag{7.17}$$

$$= \int_0^\infty (-1)^n x^n e^{-sx} f(x)\, dx = (-1)^n \mathscr{L}\{x^n f(x)\}. \tag{7.18}$$

Hence (7.15) follows.

3. *The Convolution Theorem*

This states that if

$$\mathscr{L}\{f(x)\} = \bar{f}(s) \tag{7.19}$$

and

$$\mathscr{L}\{g(x)\} = \bar{g}(s) \tag{7.20}$$

then

$$\mathscr{L}\left\{\int_0^x f(x-u)g(u)\, du\right\} = \bar{f}(s)\bar{g}(s). \tag{7.21}$$

(The integral $\int_0^x f(x-u)g(u)\, du$ is usually called the convolution of f and g and is denoted by $f*g$. Clearly, from (7.21), $\mathscr{L}\{f*g\} = \mathscr{L}\{g*f\}$.)

Proof Using (7.8), (7.21) takes the form

$$\mathscr{L}\left\{\int_0^x f(x-u)g(u)\, du\right\} = \int_0^\infty e^{-sx}\, dx \int_0^x f(x-u)g(u)\, du, \tag{7.22}$$

where the region of integration in the (u, x) plane is shown in Figure 7.1. The integration in (7.22) is first performed with respect to u from $u = 0$ to $u = x$ up the vertical strip and then from $x = 0$ to ∞ by moving the vertical strip from $x = 0$ outwards to cover the whole (infinite) shaded region.

We now change the order of integration so that we integrate first along the horizontal strip from $x = u$ to ∞ and then from $u = 0$ to ∞ by moving the horizontal strip vertically from $u = 0$ upwards. Then (7.22) becomes

$$\int_0^\infty g(u)\, du \int_{x=u}^\infty e^{-xs} f(x-u)\, dx. \tag{7.23}$$

In the integral farthest to the right in (7.23), u may be regarded as a constant. Hence putting $x - u = t$, so that $dx = dt$, (7.23) becomes

$$\int_0^\infty g(u)\,du \int_{t=0}^\infty e^{-s(u+t)} f(t)\,dt. \tag{7.24}$$

Equation (7.24) can be rewritten as

$$\int_0^\infty g(u)e^{-su}\,du \int_0^\infty f(t)e^{-st}\,dt \tag{7.25}$$

simply by collecting the terms involving u together. Each integral in (7.25) is recognizable as the integral defining the Laplace transform. Hence

$$\mathscr{L}\left\{\int_0^\infty f(x-u)g(u)\,du\right\} = \mathscr{L}\{f(x)\}\mathscr{L}\{g(x)\} = \bar{f}(s)\bar{g}(s). \tag{7.26}$$

We now give three examples of the use of these theorems.

Example 1 Since

$$\mathscr{L}\{\sin(2x)\} = \frac{2}{s^2+4}, \tag{7.27}$$

using the standard result of Table 7.1, then the Shift Theorem states that

$$\mathscr{L}\{e^{-ax}\sin(2x)\} = \frac{2}{(s+a)^2+4}. \quad ◢ \tag{7.28}$$

Example 2 Since

$$\mathscr{L}\{e^{-2x}\} = \frac{1}{s+2} \tag{7.29}$$

Figure 7.1

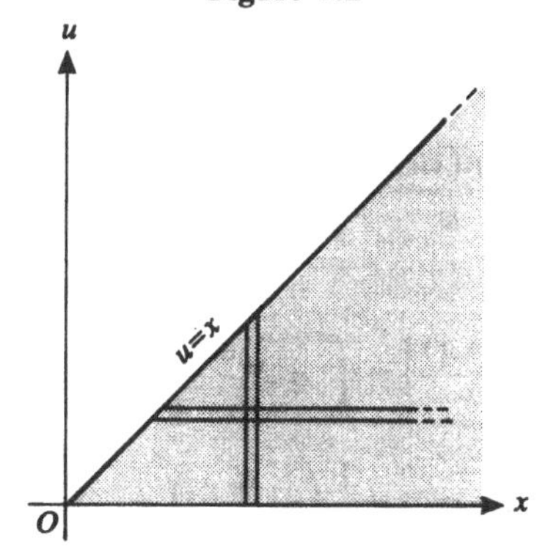

from Table 7.1, then the second theorem above states that

$$\mathscr{L}\{x\mathrm{e}^{-2x}\} = -\frac{d}{ds}\left(\frac{1}{s+2}\right) = \frac{1}{(s+2)^2} \tag{7.30}$$

and

$$\mathscr{L}\{x^2\mathrm{e}^{-2x}\} = \frac{d^2}{ds^2}\left(\frac{1}{s+2}\right) = \frac{2}{(s+2)^3}. \blacktriangleleft \tag{7.31}$$

Example 3 Given

$$I = \int_0^x \mathrm{e}^{au} \cos[b(x-u)]\, du, \tag{7.32}$$

we write $f(x-u) = \cos[b(x-u)]$ and $g(u) = \mathrm{e}^{au}$. Then $f(x) = \cos(bx)$ and $g(x) = \mathrm{e}^{ax}$ so that

$$\bar{f}(s) = \frac{s}{s^2+b^2}, \quad \bar{g}(s) = \frac{1}{s-a}. \tag{7.33}$$

Hence by the convolution theorem

$$\mathscr{L}\left\{\int_0^x \mathrm{e}^{au} \cos[b(x-u)]\, du\right\} = \bar{f}(s)\bar{g}(s) = \frac{s}{(s-a)(s^2+b^2)}. \blacktriangleleft \tag{7.34}$$

7.4 The calculation of an integral

A slightly more sophisticated problem rests on the result (which we assume without proof)

$$\mathscr{L}\left\{\int_0^\infty f(x, u)\, du\right\} = \int_0^\infty \mathscr{L}\{f(x, u)\}\, du, \tag{7.35}$$

where $\mathscr{L}$ is taken with respect to the parameter x ($x > 0$). Consider

$$\mathscr{L}\left\{\int_0^\infty \frac{\sin(xu)}{u}\, du\right\} = \int_0^\infty \mathscr{L}\left\{\frac{\sin(xu)}{u}\right\} du. \tag{7.36}$$

Now

$$\mathscr{L}\left\{\frac{\sin(xu)}{u}\right\} = \int_0^\infty \mathrm{e}^{-sx} \frac{\sin(xu)}{u}\, dx \tag{7.37}$$

$$= \frac{1}{u}\left(\frac{u}{s^2+u^2}\right) = \frac{1}{s^2+u^2}. \tag{7.38}$$

Hence

$$\mathscr{L}\left\{\int_0^\infty \frac{\sin(xu)}{u}\,du\right\} = \int_0^\infty \frac{1}{s^2+u^2}\,du \tag{7.39}$$

$$= \frac{1}{s}\left[\tan^{-1}\left(\frac{u}{s}\right)\right]_0^\infty = \frac{\pi}{2s}. \tag{7.40}$$

Now since (using Table 7.1) $\mathscr{L}\{\pi/2\} = \pi/2s$, we have, for $x > 0$,

$$\int_0^\infty \frac{\sin(xu)}{u}\,du = \frac{\pi}{2}. \tag{7.41}$$

Clearly if $x = 0$ then

$$\int_0^\infty \frac{\sin(xu)}{u}\,du = 0 \tag{7.42}$$

and if $x < 0$, from (7.41),

$$\int_0^\infty \frac{\sin(xu)}{u}\,du = -\frac{\pi}{2}. \tag{7.43}$$

A convenient way to represent the results (7.41)–(7.43) is by defining the *sign* function, $\operatorname{sgn} x$, as

$$\operatorname{sgn} x = \begin{cases} 1, & x > 0 \\ 0, & x = 0 \\ -1, & x < 0. \end{cases} \tag{7.44}$$

Then

$$\int_0^\infty \frac{\sin(xu)}{u}\,du = \frac{\pi}{2}\operatorname{sgn} x. \tag{7.45}$$

The Laplace transform approach to integration is a useful technique and may be applied to more complicated integrals.

7.5 Laplace transform of an error function

In Chapter 8 we shall require the Laplace transform of a complementary error function defined in (2.49) by

$$\operatorname{erfc} y = 1 - \operatorname{erf} y = \frac{2}{\sqrt{\pi}}\int_y^\infty \mathrm{e}^{-u^2}\,du. \tag{7.46}$$

The particular transform needed is

$$\mathscr{L}\{\operatorname{erfc}(a/2\sqrt{x})\}, \tag{7.47}$$

where $a>0$ is a real constant. Now, from (7.46) with $y=a/2\sqrt{x}$,

$$\mathscr{L}\left\{\operatorname{erfc}\left(\frac{a}{2\sqrt{x}}\right)\right\}=\frac{2}{\sqrt{\pi}}\int_0^\infty e^{-sx}\left(\int_{a/2\sqrt{x}}^\infty e^{-u^2}\,du\right)dx. \tag{7.48}$$

The region of integration in the (u, x) plane is shown in Figure 7.2. Instead of integrating with respect to u first (along the vertical strip), we change the order of integration and first integrate with respect to x (along the horizontal strip). Then (7.48) becomes

$$\mathscr{L}\left\{\operatorname{erfc}\left(\frac{a}{2\sqrt{x}}\right)\right\}=\frac{2}{\sqrt{\pi}}\int_{u=0}^\infty e^{-u^2}\left(\int_{x=a^2/4u^2}^\infty e^{-sx}\,dx\right)du \tag{7.49}$$

$$=\frac{2}{s\sqrt{\pi}}\int_0^\infty e^{-u^2}e^{-a^2s/4u^2}\,du \tag{7.50}$$

$$=\frac{2}{s\sqrt{\pi}}\int_0^\infty e^{-(u^2+a^2s/4u^2)}\,du. \tag{7.51}$$

Now writing

$$u^2+\frac{a^2s}{4u^2}=(u-a\sqrt{s}/2u)^2+a\sqrt{s} \tag{7.52}$$

and making the substitution

$$v=u-a\sqrt{s}/2u, \tag{7.53}$$

we find

$$u=\tfrac{1}{2}[v+\sqrt{(v^2+2a\sqrt{s})}]. \tag{7.54}$$

and

$$du=\tfrac{1}{2}dv+\frac{v\,dv}{\sqrt{(v^2+2a\sqrt{s})}}. \tag{7.55}$$

Figure 7.2

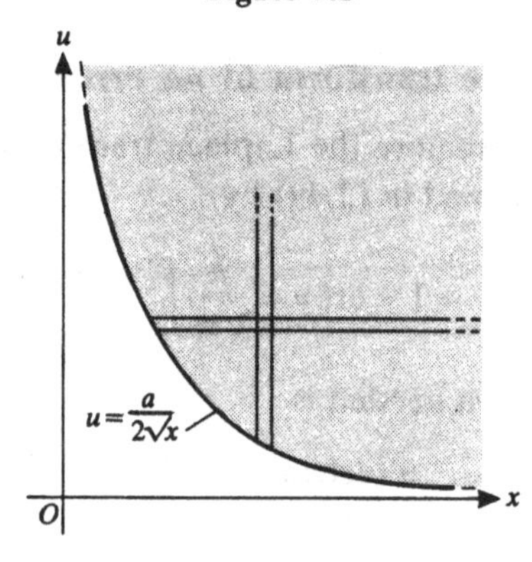

The range $0 \leqslant u < \infty$ becomes $-\infty < v < \infty$ (from (7.53)). Hence the integral in (7.51) transforms to

$$\frac{e^{-a\sqrt{s}}}{s\sqrt{\pi}}\left[\int_{-\infty}^{\infty} e^{-v^2}\, dv + \int_{-\infty}^{\infty} \frac{ve^{-v^2}}{\sqrt{(v^2+2a\sqrt{s})}}\, dv\right]. \tag{7.56}$$

The first integral has the value $\sqrt{\pi}$ (see (6.45)), and the second integral is zero since the integrand is an odd function of v. Accordingly from (7.56)

$$\mathscr{L}\left\{\operatorname{erfc}\left(\frac{a}{2\sqrt{x}}\right)\right\} = \frac{e^{-a\sqrt{s}}}{s}. \tag{7.57}$$

7.6 Transforms of the Heaviside step function and the Dirac delta-function

The Heaviside unit step function situated at $x = a$ is defined by

$$H(x-a) = \begin{cases} 1, & x > a \\ 0, & x < a. \end{cases} \tag{7.58}$$

This is a suitable (discontinuous) function for describing an 'off–on' process in physical modelling and is shown in Figure 7.3. If x is a time variable then prior to $x = a$ the function is zero, whereas after $x = a$ the function has a non-zero (unit) effect. Due to its occurrence in this way, we require its Laplace transform (see Example 10 of this chapter). Consider

$$\mathscr{L}\{H(x-a)\} = \int_0^{\infty} H(x-a)e^{-sx}\, dx \tag{7.59}$$

$$= \int_0^{a} H(x-a)e^{-sx}\, dx + \int_a^{\infty} H(x-a)e^{-sx}\, dx \tag{7.60}$$

$$= 0 + \int_a^{\infty} e^{-sx}\, dx = e^{-sa}/s, \tag{7.61}$$

using the definition (7.58).

Figure 7.3

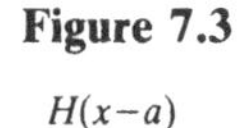

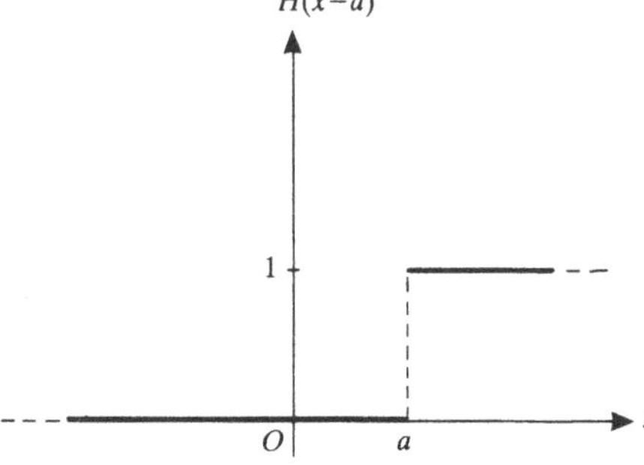

Similarly, if y is some arbitrary (Laplace transformable) function then

$$\mathscr{L}\{H(x-a)y(x-a)\} = \int_0^\infty e^{-sx}H(x-a)y(x-a)\,dx \tag{7.62}$$

$$= \int_0^a e^{-sx}H(x-a)y(x-a)\,dx + \int_a^\infty e^{-sx}H(x-a)y(x-a)\,dx \tag{7.63}$$

$$= 0 + \int_a^\infty e^{-sx}y(x-a)\,dx, \tag{7.64}$$

again using (7.58). Now writing $x - a = v$, the integral (7.64) becomes

$$\int_0^\infty e^{-s(v+a)}y(v)\,dv = e^{-sa}\bar{y}(s), \tag{7.65}$$

where $\bar{y}(s)$ is the Laplace transform of $y(x)$. Hence

$$\mathscr{L}\{H(x-a)y(x-a)\} = e^{-sa}\bar{y}(s). \tag{7.66}$$

The Dirac delta-function is defined by $\delta(x-a) = 0$, $x \neq a$, and

$$\int_{-\infty}^\infty \delta(x-a)\,dx = 1, \tag{7.67}$$

$$\int_{-\infty}^\infty \delta(x-a)f(x)\,dx = f(a). \tag{7.68}$$

The delta-function may be considered as the limit of a sequence of ordinary functions. For example, defining

$$\delta_n(x) = \sqrt{(n/\pi)}e^{-nx^2} \quad (n = 1, 2, 3, \ldots), \tag{7.69}$$

then

$$\int_{-\infty}^\infty \sqrt{(n/\pi)}e^{-nx^2}\,dx = 1, \tag{7.70}$$

using the standard integral $\int_{-\infty}^\infty e^{-nx^2}\,dx = \sqrt{(\pi/n)}$. Hence

$$\lim_{n\to\infty}\int_{-\infty}^\infty \delta_n(x)\,dx = \int_{-\infty}^\infty \delta(x)\,dx = 1, \tag{7.71}$$

and the delta-function is seen to be the limit of a sequence of Gaussian functions with decreasing widths.

If we consider the integral of $\delta(x-a)$ over the interval $-\infty$ to x', then

$$\int_{-\infty}^{x'} \delta(x-a)\,dx = \begin{cases} 0, & \text{if } x' < a \\ 1, & \text{if } x' > a \end{cases} \tag{7.72}$$

$$= H(x'-a), \tag{7.73}$$

using (7.58). Hence by *formally* differentiating each side we obtain

$$\delta(x-a) = \frac{d}{dx} H(x-a). \tag{7.74}$$

This shows that the delta-function is not a proper function in the usual sense since the right-hand side of (7.74) is the 'differential' of a discontinuous function.

The Laplace transform of the delta-function is easily obtained as follows:

$$\mathscr{L}\{\delta(x-a)\} = \int_0^\infty \mathrm{e}^{-sx}\delta(x-a)\,dx = \mathrm{e}^{-sa}, \quad a > 0, \tag{7.75}$$

using the property (7.68).

7.7 Transforms of derivatives

One of the main applications of the Laplace transform is to the solution of differential equations, both ordinary and partial, and we therefore require the transforms of the derivatives of the dependent variable. For the purposes of this discussion, we do not go beyond the second differential coefficient.

1. *Ordinary derivatives*

Suppose $y = y(x)$ is a Laplace transformable function. Then

$$\mathscr{L}\left\{\frac{dy}{dx}\right\} = \int_0^\infty \mathrm{e}^{-sx}\frac{dy}{dx}\,dx = \left[y\mathrm{e}^{-sx}\right]_0^\infty + s\int_0^\infty y\mathrm{e}^{-sx}\,dx \tag{7.76}$$

$$= -y(0) + s\bar{y}(s), \tag{7.77}$$

where $\bar{y}(s)$ is the Laplace transform of $y(x)$ and $y(0)$ is the value of y at $x = 0$. We have assumed here that $y\mathrm{e}^{-sx} \to 0$ as $x \to \infty$. Similarly

$$\mathscr{L}\left\{\frac{d^2y}{dx^2}\right\} = \int_0^\infty \mathrm{e}^{-sx}\frac{d^2y}{dx^2}\,dx = \left[\mathrm{e}^{-sx}\frac{dy}{dx}\right]_0^\infty + s\int_0^\infty \mathrm{e}^{-sx}\frac{dy}{dx}\,dx \tag{7.78}$$

$$= -y'(0) + s\mathscr{L}\left\{\frac{dy}{dx}\right\} \tag{7.79}$$

$$= -y'(0) + s[s\bar{y}(s) - y(0)], \tag{7.80}$$

Table 7.2

$f(x)$	$\mathscr{L}\{f(x)\}=\bar{f}(s)$
$\dfrac{dy}{dx}$	$s\bar{y}(s)-y(0)$
$\dfrac{d^2y}{dx^2}$	$s^2\bar{y}(s)-sy(0)-y'(0)$
$x\dfrac{dy}{dx}$	$-s\dfrac{d\bar{y}(s)}{ds}-\bar{y}(s)$
$x^2\dfrac{dy}{dx}$	$s\dfrac{d^2\bar{y}(s)}{ds^2}+2\dfrac{d\bar{y}(s)}{ds}$
$x\dfrac{d^2y}{dx^2}$	$-s^2\dfrac{d\bar{y}(s)}{ds}-2s\bar{y}(s)+y(0)$
$x^2\dfrac{d^2y}{dx^2}$	$s^2\dfrac{d^2\bar{y}(s)}{ds^2}+4s\dfrac{d\bar{y}(s)}{ds}+2\bar{y}(s)$

using (7.77), where we have assumed $e^{-sx}y'(x)\to 0$ as $x\to\infty$. Hence

$$\mathscr{L}\left\{\frac{d^2y}{dx^2}\right\}=s^2\bar{y}(s)-sy(0)-y'(0), \tag{7.81}$$

where $y'(0)$ is the value of dy/dx at $x=0$.

The transforms of terms of the form $x^m\,d^ny/dx^n$ may be obtained using (7.15). For example

$$\mathscr{L}\left\{x\frac{dy}{dx}\right\}=-\frac{d}{ds}\left[\mathscr{L}\left\{\frac{dy}{dx}\right\}\right]=-\frac{d}{ds}[s\bar{y}(s)-y(0)] \tag{7.82}$$

$$=-s\frac{d\bar{y}(s)}{ds}-\bar{y}(s). \tag{7.83}$$

Some of these results are collected together in Table 7.2.

2. *Partial derivatives*

We now suppose that $u(x, t)$ is an arbitrary function of x and t, where $a\leqslant x\leqslant b$ and $t\geqslant 0$, a and b being constants. Since t ranges from 0 to ∞, we may transform with respect to this variable as follows:

$$\mathscr{L}\left\{\frac{\partial}{\partial t}u(x, t)\right\}=\int_0^\infty e^{-st}\frac{\partial u(x, t)}{\partial t}\,dt \tag{7.84}$$

$$=[u(x, t)e^{-st}]_0^\infty+s\int_0^\infty u(x, t)e^{-st}\,dt \tag{7.85}$$

$$=s\bar{u}(x, s)-u(x, 0), \tag{7.86}$$

where $\bar{u}(x, s)$ is the Laplace transform of $u(x, t)$ with respect to t. Further

$$\mathscr{L}\left\{\frac{\partial^2}{\partial t^2} u(x, t)\right\} = \int_0^\infty e^{-st} \frac{\partial^2 u(x, t)}{\partial t^2}\, dt$$

$$= s^2\bar{u}(x, s) - su(x, 0) - \left.\frac{\partial u(x, t)}{\partial t}\right|_{t=0}. \tag{7.87}$$

Again we have assumed that u and its first derivative (with respect to t) are of exponential order and are such that at infinity the integrated terms are zero.

Besides the derivatives $\partial u/\partial t$ and $\partial^2 u/\partial t^2$, we have the other derivatives $\partial u/\partial x$, $\partial^2 u/\partial x^2$, and $\partial^2 u/\partial x\, \partial t$. For the first of these, we have

$$\mathscr{L}\left\{\frac{\partial}{\partial x} u(x, t)\right\} = \int_0^\infty e^{-st} \frac{\partial u(x, t)}{\partial x}\, dt = \frac{d}{dx}\int_0^\infty u(x, t) e^{-st}\, dt \tag{7.88}$$

$$= \frac{d}{dx}\bar{u}(x, s), \tag{7.89}$$

where s is treated as a parameter. Similarly

$$\mathscr{L}\left\{\frac{\partial^2 u}{\partial x^2}\right\} = \frac{d^2\bar{u}(x, s)}{dx^2} \tag{7.90}$$

and

$$\mathscr{L}\left\{\frac{\partial^2 u}{\partial x\, \partial t}\right\} = \frac{d}{dx}\mathscr{L}\left\{\frac{\partial u}{\partial t}\right\} = \frac{d}{dx}[s\bar{u}(x, s) - u(x, 0)]. \tag{7.91}$$

7.8 Inversion

Before applying the Laplace transform to the solution of differential equations, it is necessary to discuss in detail how to find $f(x)$ from the transform $\bar{f}(s)$. Such a process was indicated in (7.7) but we now need specific techniques. One of the simplest and most obvious is to read the inversion from a list of transforms so that from Table 7.1, for example,

$$\mathscr{L}^{-1}\left\{\frac{s}{s^2 + a^2}\right\} = \cos(ax). \tag{7.92}$$

Other methods use partial fractions, the Convolution Theorem, and the general inversion formula based on contour integration.

1. *Partial fraction method*

When the function $\bar{f}(s)$ has the form

$$\bar{f}(s) = P(s)/Q(s), \tag{7.93}$$

where $P(s)$ and $Q(s)$ are polynomials in s, the degree of P being less than that of Q, the method of partial fractions may be used to express $\bar{f}(s)$ as the sum of terms the inversions of which are readily found.

Example 4 Consider

$$\bar{f}(s) = \frac{2s^2 + 3s - 4}{(s-2)(s^2+2s+2)}. \tag{7.94}$$

Then by partial fractions we may write this as

$$\bar{f}(s) = \frac{1}{s-2} + \frac{s+3}{(s+1)^2+1} \tag{7.95}$$

$$= \frac{1}{s-2} + \frac{s+1}{(s+1)^2+1} + \frac{2}{(s+1)^2+1}. \tag{7.96}$$

Hence

$$f(x) = \mathscr{L}^{-1}\{\bar{f}(s)\} = \mathscr{L}^{-1}\left\{\frac{1}{s-2}\right\} + \mathscr{L}^{-1}\left\{\frac{s+1}{(s+1)^2+1}\right\}$$

$$+ \mathscr{L}^{-1}\left\{\frac{2}{(s+1)^2+1}\right\} \tag{7.97}$$

$$= e^{2x} + e^{-x}\cos x + 2e^{-x}\sin x, \tag{7.98}$$

using the standard results of Table 7.1 and the Shift Theorem. ◢

2. *Inversion using the Convolution Theorem*

From (7.26) we may write

$$\mathscr{L}^{-1}\{\bar{f}(s)\bar{g}(s)\} = \int_0^x f(x-u)g(u)\,du. \tag{7.99}$$

Hence if some function, say $\bar{F}(s)$, is given which can be written as the product of two functions $\bar{f}(s)$ and $\bar{g}(s)$, the inversions of which are known, we may then use (7.99) to obtain the inversion of $\bar{F}(s)$.

Example 5 To evaluate

$$\mathscr{L}^{-1}\left\{\frac{4}{s^2(s+2)^2}\right\}, \tag{7.100}$$

we write

$$\bar{f}(s) = \frac{4}{s^2}, \quad \bar{g}(s) = \frac{1}{(s+2)^2}. \tag{7.101}$$

Then from the standard results and the Shift Theorem we find

$$f(x) = \mathscr{L}^{-1}\{\bar{f}(s)\} = 4x, \tag{7.102}$$

$$g(x) = \mathscr{L}^{-1}\{\bar{g}(s)\} = x\mathrm{e}^{-2x}. \tag{7.103}$$

Hence from (7.99), since $f(x-u) = 4(x-u)$ and $g(u) = u\mathrm{e}^{-2u}$, we have

$$\mathscr{L}^{-1}\left\{\frac{4}{s^2(s+2)^2}\right\} = 4\int_0^x (x-u)u\mathrm{e}^{-2u}\,du, \tag{7.104}$$

which, by integration by parts, gives

$$\mathscr{L}^{-1}\left\{\frac{4}{s^2(s+2)^2}\right\} = (x+1)\mathrm{e}^{-2x} + x - 1. \quad ◢ \tag{7.105}$$

Example 6 Given that

$$f(x) = \mathscr{L}^{-1}\{\bar{f}(s)\}, \tag{7.106}$$

we can show that

$$\mathscr{L}^{-1}\left\{\frac{1}{s}\bar{f}(s)\right\} = \int_0^x f(t)\,dt. \tag{7.107}$$

Writing

$$\bar{g}(s) = 1/s, \tag{7.108}$$

we have that $g(x) = 1$. Hence, using the Convolution Theorem,

$$\mathscr{L}^{-1}\left\{\frac{1}{s}\bar{f}(s)\right\} = \int_0^x f(x-u)g(u)\,du \tag{7.109}$$

$$= \int_0^x f(x-u)\,du, \tag{7.110}$$

since $g(u) = 1$. Writing $x - u = t$, we have finally

$$\mathscr{L}^{-1}\left\{\frac{1}{s}\bar{f}(s)\right\} = \int_0^x f(t)\,dt. \quad ◢ \tag{7.111}$$

3. Inversion by contour integration: the general inversion formula

It may be shown in general that

$$f(x) = \frac{1}{2\pi \mathrm{i}}\int_{\gamma - \mathrm{i}\infty}^{\gamma + \mathrm{i}\infty} \mathrm{e}^{sx}\bar{f}(s)\,ds, \tag{7.112}$$

where s is now a *complex* variable. The path of integration is the straight line L in the complex s-plane with equation $s = \gamma + iv$, $-\infty < v < \infty$, as shown in Figure 7.4, γ being chosen so that all the singularities of the integrand lie to the left of this line. This line is known as the Bromwich contour. In practice, we evaluate the integral in (7.112) by taking a finite straight line, closing the contour in a suitable way and then letting the closed contour expand to infinity, while using the Cauchy Residue Theorem. There are a number of ways of closing the contour. Here we concentrate on the method shown in Figure 7.5 in which the finite straight line L' is closed by means of the part of the circle C, centre the origin. Then, provided that $|\bar{f}(s)| < \alpha R^{-k}$ on $s = Re^{i\theta}$, where α and k are positive constants, the integral of $e^{sx}\bar{f}(s)$ on C tends to zero as $R \to \infty$, as we now show.

Consider first the part of C which is a semicircle S to the left of the

Figure 7.4

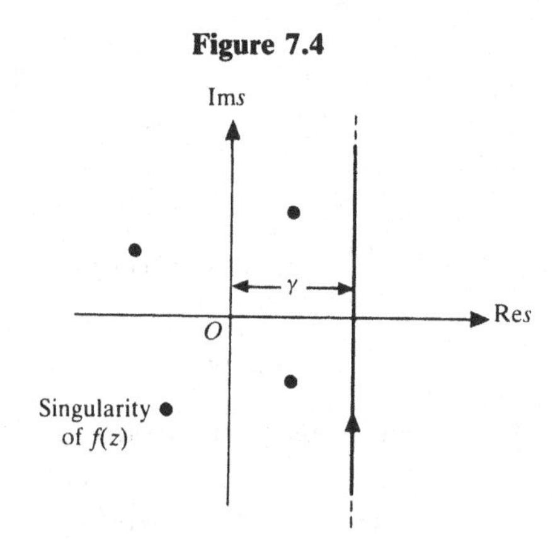

Figure 7.5

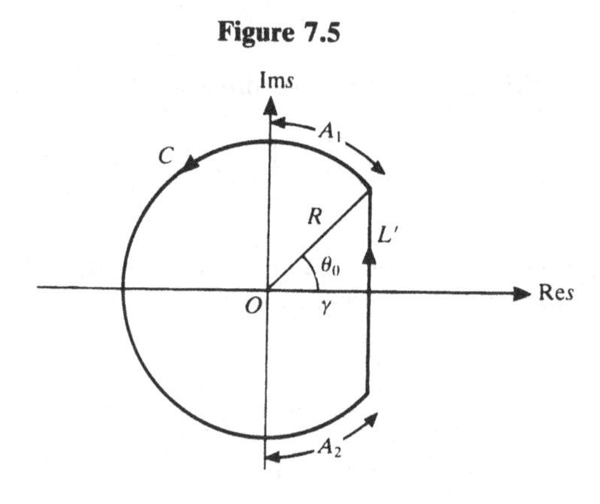

y-axis. On this part $s = Re^{i\theta}$, where $\pi/2 < \theta < 3\pi/2$. Hence on S

$$|e^{sx}\bar{f}(s)\,ds| = |e^{Rx(\cos\theta + i\sin\theta)}\bar{f}(Re^{i\theta})Re^{i\theta}i\,d\theta|$$
$$= e^{Rx\cos\theta}\,|\bar{f}(Re^{i\theta})|\,R\,d\theta. \tag{7.113}$$

By assumption $\bar{f}$ behaves as a power of R on this semicircle and $\cos\theta < 0$. Hence as $R \to \infty$, (7.113) tends to zero since it includes a decreasing exponential (with R) multiplied by a power of R.

Now consider the upper arc of the circle A_1 in the first quadrant. The contribution from this arc to the total integral is bounded as follows:

$$\left|\int_{A_1} e^{sx}\bar{f}(s)\,ds\right| < \int_{A_1} |e^{sx}\bar{f}(s)\,ds| < \int_{\theta_0}^{\pi/2} e^{Rx\cos\theta}\frac{\alpha}{R^k}R\,d\theta, \tag{7.114}$$

where $\cos\theta_0 = \gamma/R$ (see Figure 7.5). Now since $\theta_0 \leqslant \theta \leqslant \pi/2$, it follows that $\cos\theta \leqslant \cos\theta_0 = \gamma/R$. Hence (7.114) is less than

$$\frac{\alpha}{R^{k-1}}\int_{\theta_0}^{\pi/2} e^{Rx\gamma/R}\,d\theta = \frac{\alpha e^{\gamma x}}{R^{k-1}}(\pi/2 - \theta_0). \tag{7.115}$$

In (7.115), the quantity $\pi/2 - \theta_0 = \pi/2 - \cos^{-1}(\gamma/R)$. For small t, the function $y = \cos^{-1} t$ has the expansion

$$y = y(0) + \frac{t}{1!}y'(0) + \ldots = \frac{\pi}{2} - t + \ldots. \tag{7.116}$$

Hence for large R, $t = \gamma/R$ is small and (7.116) gives

$$\frac{\pi}{2} - \cos^{-1}\left(\frac{\gamma}{R}\right) = \frac{\pi}{2} - \left(\frac{\pi}{2} - \frac{\gamma}{R} + \ldots\right) = \frac{\gamma}{R} + \ldots. \tag{7.117}$$

Finally then, for large R, (7.115) behaves as

$$\frac{\alpha e^{\gamma x}}{R^{k-1}}\frac{\gamma}{R} = \alpha\gamma\frac{e^{\gamma x}}{R^k}, \tag{7.118}$$

which tends to zero as $R \to \infty$ since $k > 0$.

Similarly the integral on the lower arc A_2 in the fourth quadrant tends to zero as $R \to \infty$. By the Cauchy Residue Theorem we therefore have

$$\int_S e^{sx}\bar{f}(s)\,ds + \int_{A_1} e^{sx}\bar{f}(s)\,ds + \int_{A_2} e^{sx}\bar{f}(s)\,ds + \int_{L'} e^{sx}\bar{f}(s)\,ds$$
$$= 2\pi i \times [\text{sum of the residues of } e^{sx}\bar{f}(s) \text{ inside the contour}]. \tag{7.119}$$

Letting $R \to \infty$, the first three integrals in (7.119) tend to zero and the finite line L' becomes the Bromwich contour L. Further L is chosen to lie to the right of all the singularities of $\bar{f}(s)$. Hence

$$f(x) = \frac{1}{2\pi \mathrm{i}} \int_L \mathrm{e}^{sx} \bar{f}(s)\, ds$$
$$= [\text{sum of the residues of } \mathrm{e}^{sx}\bar{f}(s) \text{ at the poles of } \bar{f}(s)], \quad (7.120)$$

since e^{sx} has no poles.

Example 7 Consider the function

$$\bar{f}(s) = \frac{s}{s^2+4} \quad (7.121)$$

which has simple poles at $s = \pm 2\mathrm{i}$. Using the formula for the residue at a simple pole (5.183), we have

$$(\text{residue at } s = 2\mathrm{i}) = \lim_{s \to 2\mathrm{i}} \left[(s - 2\mathrm{i}) \frac{s\mathrm{e}^{sx}}{(s-2\mathrm{i})(s+2\mathrm{i})} \right] = \frac{\mathrm{e}^{2\mathrm{i}x}}{2} \quad (7.122)$$

and

$$(\text{residue at } s = -2\mathrm{i}) = \lim_{s \to -2\mathrm{i}} \left[(s + 2\mathrm{i}) \frac{s\mathrm{e}^{sx}}{(s-2\mathrm{i})(s+2\mathrm{i})} \right] = \frac{\mathrm{e}^{-2\mathrm{i}x}}{2}. \quad (7.123)$$

Hence

$$f(x) = (\text{sum of residues}) = \tfrac{1}{2}(\mathrm{e}^{2\mathrm{i}x} + \mathrm{e}^{-2\mathrm{i}x}) = \cos(2x), \quad (7.124)$$

which is the inversion of (7.121) (as can be seen from Table 7.1). ◢

Example 8 Consider the function in Example 5

$$\bar{f}(s) = \frac{4}{s^2(s+2)^2} \quad (7.125)$$

which has double poles at $s = 0$ and $s = -2$. Using the residue formula for double poles ((5.206) with $m = 2$) gives

$$(\text{residue at } s = 0) = \lim_{s \to 0} \frac{d}{ds} \left[s^2 \frac{4\mathrm{e}^{sx}}{s^2(s+2)^2} \right] = x - 1 \quad (7.126)$$

and

$$(\text{residue at } s = -2) = \lim_{s \to -2} \frac{d}{ds} \left[(s+2)^2 \frac{4\mathrm{e}^{sx}}{s^2(s+2)^2} \right] = (x+1)\mathrm{e}^{-2x}. \quad (7.127)$$

Hence

$$f(x) = (\text{sum of residues}) = (x+1)\mathrm{e}^{-2x} + x - 1, \quad (7.128)$$

as in (7.105). ◢

7.9 Inversions of functions with branch points

When the function to be inverted possesses a branch point, we must proceed as in Section 6.5 by making a cut in the s-plane so as to exclude this point from the contour. The method of closing the Bromwich contour therefore requires modification. Consider the following example:

Example 9 Suppose we require

$$\mathscr{L}^{-1}\{1/\sqrt{s}\}. \tag{7.129}$$

We first anticipate the result by evaluating the Laplace transform of $1/\sqrt{x}$ by elementary methods.

$$\mathscr{L}\left\{\frac{1}{\sqrt{x}}\right\} = \int_0^\infty \frac{1}{\sqrt{x}}\,\mathrm{e}^{-sx}\,dx \tag{7.130}$$

$$= \int_0^\infty \frac{1}{u}\,\mathrm{e}^{-su^2}2u\,du = 2\int_0^\infty \mathrm{e}^{-su^2}\,du = \sqrt{\left(\frac{\pi}{s}\right)}. \tag{7.131}$$

Hence

$$\mathscr{L}^{-1}\{1/\sqrt{s}\} = 1/\sqrt{(\pi x)}. \tag{7.132}$$

Now $\bar{f}(s) = 1/\sqrt{s}$ has a branch point at $s = 0$. We can allow for this by cutting the circular closure of the Bromwich contour as shown in Figure 7.6 and surrounding the branch point at $s = 0$ by a small circle C' (radius ρ). Denoting the large curved part of radius R by C and the two straight parts defining the cut by L_1 and L_2, then

$$\frac{1}{2\pi\mathrm{i}}\left\{\int_{L'} + \int_C + \int_{L_1} + \int_{L_2} + \int_{C'}\right\} = 0, \tag{7.133}$$

Figure 7.6

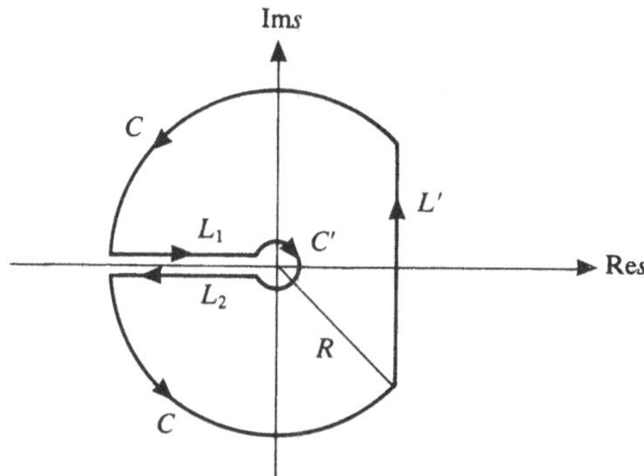

since $1/\sqrt{s}$ has no singularities inside the total contour. We showed in the last section that the integral on C tends to zero as $R \to \infty$, whilst that on L' tends to the required Bromwich integral. We now examine the remaining three integrals in (7.133).

On C', we have $s = \rho e^{i\theta}$ and

$$\int_{C'} \left| \frac{1}{\sqrt{s}} e^{sx} \, ds \right| = \int_{\pi}^{-\pi} \left| \frac{e^{x\rho(\cos\theta + i\sin\theta)}}{\sqrt{\rho}\, e^{i\theta/2}} \rho i e^{i\theta} \, d\theta \right|$$

$$= \int_{\pi}^{-\pi} \sqrt{\rho}\, e^{x\rho\cos\theta} \, d\theta, \tag{7.134}$$

which tends to zero as $\rho \to 0$.

On L_1, $s = re^{i\pi} = -r$ so that $\sqrt{s} = \sqrt{r}\, e^{i\pi/2} = i\sqrt{r}$. Hence

$$\int_{L_1} \frac{e^{sx}}{\sqrt{s}} ds = \int_{-\infty}^{0} \frac{e^{sx}}{\sqrt{s}} ds = \int_{\infty}^{0} \frac{e^{-rx}}{i\sqrt{r}} (-dr) = -i \int_0^{\infty} \frac{e^{-rx}}{\sqrt{r}} dr. \tag{7.135}$$

On L_2, $s = re^{-i\pi} = -r$ so that $\sqrt{s} = \sqrt{r}\, e^{-i\pi/2} = -i\sqrt{r}$. Hence

$$\int_{L_2} \frac{e^{sx}}{\sqrt{s}} ds = \int_{0}^{-\infty} \frac{e^{sx}}{\sqrt{s}} ds = \int_{0}^{\infty} \frac{e^{-rx}}{-i\sqrt{r}} (-dr) = -i \int_0^{\infty} \frac{e^{-rx}}{\sqrt{r}} dr. \tag{7.136}$$

The cut constrains the argument of s to lie between $-\pi$ and π giving an argument of π above the cut and $-\pi$ below it. Consequently using (7.133)–(7.136) we have, as $R \to \infty$ and $\rho \to 0$,

$$f(x) = \frac{1}{2\pi i} \int_{\gamma - i\infty}^{\gamma + i\infty} \frac{e^{sx}}{\sqrt{s}} ds = -\frac{1}{2\pi i} 2(-i) \int_0^{\infty} \frac{e^{-rx}}{\sqrt{r}} dr \tag{7.137}$$

$$= \frac{1}{\pi} \int_0^{\infty} \frac{e^{-rx}}{\sqrt{r}} dr. \tag{7.138}$$

Putting $rx = u^2$, say, (7.138) becomes

$$f(x) = \frac{1}{\pi} \int_0^{\infty} \frac{e^{-u^2}}{(u/\sqrt{x})} \frac{2u}{x} du = \frac{2}{\pi\sqrt{x}} \int_0^{\infty} e^{-u^2} du = \frac{1}{\sqrt{(\pi x)}}, \tag{7.139}$$

as found in (7.132). ◢

The following two results have occurred previously and may be obtained by the contour integration method:

(1)
$$\mathscr{L}^{-1}\left\{\frac{e^{-a\sqrt{s}}}{s}\right\} = 1 - \operatorname{erf}\left(\frac{a}{2\sqrt{x}}\right) = \operatorname{erfc}\left(\frac{a}{2\sqrt{x}}\right), \tag{7.140}$$

(see (7.57)).

(2)
$$\mathscr{L}^{-1}\left\{\frac{1}{s}\ln\left(\frac{s+a}{a}\right)\right\} = E_1(at), \quad a > 0, \tag{7.141}$$

(see (2.45)).

7.10 Solution of ordinary differential equations

It is often convenient to use Laplace transforms to solve ordinary differential equations using the results of Table 7.2.

Example 10 Consider

$$\frac{d^2y}{dx^2}+3\frac{dy}{dx}+2y=2-2H(x-1), \tag{7.142}$$

given that $y(0)=0$ and $y'(0)=2$.

Taking the Laplace transform of the equation term by term and using the results (7.61), (7.77), and (7.80), we obtain

$$s^2\bar{y}(s)-sy(0)-y'(0)+3[s\bar{y}(s)-y(0)]+2\bar{y}(s)=\frac{2}{s}-\frac{2e^{-s}}{s}. \tag{7.143}$$

Hence

$$(s^2+3s+2)\bar{y}(s)-2=\frac{2}{s}-\frac{2e^{-s}}{s} \tag{7.144}$$

so that

$$\bar{y}(s)=\frac{1}{(s^2+3s+2)}\left[\frac{2(s+1)}{s}-\frac{2e^{-s}}{s}\right] \tag{7.145}$$

$$=\frac{2}{s(s+2)}-\frac{2e^{-s}}{s(s+1)(s+2)}. \tag{7.146}$$

Writing this expression in partial fractions, we find

$$\bar{y}(s)=\frac{1}{s}-\frac{1}{(s+2)}-e^{-s}\left[\frac{1}{s}-\frac{2}{(s+1)}+\frac{1}{(s+2)}\right]. \tag{7.147}$$

Inverting (7.147) using the Shift Theorem (7.11) and result (7.66), we have

$$y(x)=1-e^{-2x}+H(x-1)[-1+2e^{-(x-1)}-e^{-2(x-1)}]. \blacktriangleleft \tag{7.148}$$

The attraction of this method is that it enables us to solve the differential equation by solving only an algebraic equation (7.144) and using known results for Laplace inversions. It is sometimes convenient to use the differential equation defining a special function to evaluate its Laplace transform. Consider the following example:

Example 11 Suppose we require the Laplace transform of the Bessel function of order zero, $J_0(x)$. The differential equation satisfied by $y=J_0(x)$ is (see (2.61) with $\nu=0$)

$$x\frac{d^2y}{dx^2}+\frac{dy}{dx}+xy=0. \tag{7.149}$$

Taking the Laplace transform of this equation, using Table 7.2, we obtain

$$\left[-s^2\frac{d\bar{y}}{ds}-2s\bar{y}+y(0)\right]+[s\bar{y}-y(0)]-\frac{d\bar{y}}{ds}=0, \tag{7.150}$$

which simplifies to

$$(s^2+1)\frac{d\bar{y}}{ds}+s\bar{y}=0. \tag{7.151}$$

Separating the variables and integrating gives

$$\bar{y}(s)=\frac{A}{\sqrt{(1+s^2)}}, \tag{7.152}$$

where A is an arbitrary constant which must be evaluated. Hence the Laplace transform of $J_0(x)$ is

$$\frac{A}{\sqrt{(1+s^2)}}=\int_0^\infty e^{-sx}J_0(x)\,dx. \tag{7.153}$$

Putting $s=0$ gives

$$A=\int_0^\infty J_0(x)\,dx=1, \tag{7.154}$$

using the result of Problem 8, Chapter 2. Hence the Laplace transform of $J_0(x)$ is $1/\sqrt{(1+s^2)}$. We note that this result could have been found by substituting the series for $J_0(x)$ into the Laplace integral and integrating term by term. Summing the resulting series gives the expansion of $1/\sqrt{(1+s^2)}$. However, using this method, we can prove convergence only for $|s|>1$. ◢

7.11 Solution of a Volterra integral equation

The equation

$$y(x)=f(x)+\int_0^x y(u)g(x-u)\,du \tag{7.155}$$

is a type of Volterra integral equation for $y(x)$, where $f(x)$ and $g(x)$ are given functions. The Laplace transform approach is particularly suited to this form of integral equation since the integral is a convolution of y and g. Hence transforming and using the Convolution Theorem (7.21), we have

$$\bar{y}(s)=\bar{f}(s)+\bar{y}(s)\bar{g}(s), \tag{7.156}$$

whence

$$\bar{y}(s)=\frac{\bar{f}(s)}{1-\bar{g}(s)}. \tag{7.157}$$

Inverting (7.157) gives $y(x)$.

Example 12 Consider the integral equation

$$y(x)=\sin x+\int_0^x \sin[2(x-u)]y(u)\,du. \tag{7.158}$$

Taking the Laplace transform gives

$$\bar{y}(s)=\frac{1}{1+s^2}+\frac{2}{4+s^2}\bar{y}(s) \tag{7.159}$$

or

$$\bar{y}(s)=\frac{(4+s^2)}{(1+s^2)(2+s^2)}=\frac{3}{1+s^2}-\frac{2}{2+s^2}. \tag{7.160}$$

Inverting gives

$$y(x)=3\sin x-\sqrt{2}\sin(\sqrt{2}\,x). \quad ◢ \tag{7.161}$$

7.12 The Fourier transform

In this section, we give the corresponding results to the Laplace transform for the Fourier transform defined by

$$\bar{f}(s)=\int_{-\infty}^{\infty} f(x)\mathrm{e}^{-\mathrm{i}sx}\,dx, \tag{7.162}$$

where the notation $\mathscr{F}\{f(x)\}=\bar{f}(s)$ denotes the Fourier transform of $f(x)$. It can be shown that the inverse transform $\mathscr{F}^{-1}\{\bar{f}(s)\}$ is given by

$$\mathscr{F}^{-1}\{\bar{f}(s)\}=f(x)=\frac{1}{2\pi}\int_{-\infty}^{\infty}\bar{f}(s)\mathrm{e}^{\mathrm{i}sx}\,ds. \tag{7.163}$$

We note that in some books the definitions differ from those above by having different multiplicative factors in front of the integrals. Any two such factors the product of which is $1/2\pi$ can be used.

1. Even functions

If $f(x)$ is an even function ($f(-x)=f(x)$ in the range $-\infty<x<\infty$) then we may define the Fourier cosine transform $\mathscr{F}_C\{f(x)\}$ as

$$\mathscr{F}_C\{f(x)\}=\bar{f}_C(s)=\int_0^{\infty} f(x)\cos(sx)\,dx \tag{7.164}$$

with the inversion

$$\mathscr{F}^{-1}\{\bar{f}_C(s)\} = f(x) = \frac{2}{\pi}\int_0^\infty \bar{f}_C(s)\cos(sx)\,ds. \qquad (7.165)$$

2. Odd functions

If $f(x)$ is an odd function ($f(-x) = -f(x)$ in the range $-\infty < x < \infty$) then we may define the Fourier sine transform $\mathscr{F}_S\{f(x)\}$ as

$$\mathscr{F}_S\{f(x)\} = \bar{f}_S(s) = \int_0^\infty f(x)\sin(sx)\,dx \qquad (7.166)$$

with the inversion

$$\mathscr{F}_S^{-1}\{\bar{f}_S(s)\} = f(x) = \frac{2}{\pi}\int_0^\infty \bar{f}_S(s)\sin(sx)\,ds. \qquad (7.167)$$

These three transforms are linear in the sense of (7.3) and (7.5). Functions defined for $x \geqslant 0$ which are neither even nor odd can be extended over the whole range by defining them to be either even or odd, as in the case of Fourier series.

Example 13 Consider the Fourier sine transform of e^{-x} for $x > 0$ (so that the function is $-e^x$ for $x < 0$, thus making it odd). Then

$$\mathscr{F}_S\{e^{-x}\} = \int_0^\infty e^{-x}\sin(sx)\,dx = \frac{s}{s^2+1}. \qquad (7.168)$$

Using the inversion result (7.167) for the Fourier sine transform, we have for $x > 0$

$$e^{-x} = \frac{2}{\pi}\int_0^\infty \frac{s}{s^2+1}\sin(sx)\,ds. \qquad (7.169)$$

Choosing some constant value of x, say $x = k$ (and calling the dummy s of integration x) we have for $k > 0$

$$\int_0^\infty \frac{x\sin(kx)}{x^2+1}\,dx = \frac{\pi}{2}e^{-k}. \qquad (7.170)$$

For $k < 0$ we have

$$\int_0^\infty \frac{x\sin(kx)}{x^2+1}\,dx = -\frac{\pi}{2}e^{k}. \qquad (7.171)$$

3. Convolution Theorem

The corresponding result to (7.21) is

$$\mathscr{F}\left\{\int_{-\infty}^{\infty} f(x-u)g(u)\,du\right\} = \bar{f}(s)\bar{g}(s), \tag{7.172}$$

where $\bar{f}(s)$ and $\bar{g}(s)$ are the Fourier transforms of $f(x)$ and $g(x)$ as in (7.162). As with the Laplace transform we denote the convolution integral by

$$\int_{-\infty}^{\infty} f(x-u)g(u)\,du = f*g \tag{7.173}$$

with the property

$$\mathscr{F}\{f*g\} = \mathscr{F}\{g*f\}. \tag{7.174}$$

4. Transforms of derivatives

(*a*) *Fourier transform* If $y = y(x)$, we may obtain the Fourier transforms of the derivatives by simple integration by parts:

$$\mathscr{F}\left\{\frac{dy}{dx}\right\} = \int_{-\infty}^{\infty} \frac{dy}{dx} \mathrm{e}^{-\mathrm{i}sx}\,dx \tag{7.175}$$

$$= [y(x)\mathrm{e}^{-\mathrm{i}sx}]_{-\infty}^{\infty} + \mathrm{i}s \int_{-\infty}^{\infty} y(x)\mathrm{e}^{-\mathrm{i}sx}\,dx \tag{7.176}$$

$$= \mathrm{i}s\bar{y}(s), \tag{7.177}$$

provided $y(x) \to 0$ as $x \to \pm\infty$.

Similarly

$$\mathscr{F}\left\{\frac{d^2y}{dx^2}\right\} = (\mathrm{i}s)^2\bar{y}(s) \tag{7.178}$$

and, in general, provided $y^{(n-1)}(x) \to 0$ as $x \to \pm\infty$,

$$\mathscr{F}\left\{\frac{d^ny}{dx^n}\right\} = (\mathrm{i}s)^n\bar{y}(s). \tag{7.179}$$

(*b*) *Fourier cosine transform* Using the definition (7.164) and integrating by parts, we have

$$\mathscr{F}_{\mathrm{C}}\left\{\frac{dy}{dx}\right\} = \int_{0}^{\infty} \frac{dy}{dx} \cos(sx)\,dx \tag{7.180}$$

$$= [y(x)\cos(sx)]_0^{\infty} + s\int_0^{\infty} y(x)\sin(sx)\,dx \tag{7.181}$$

$$= -y(0) + s\bar{y}_{\mathrm{S}}(s), \tag{7.182}$$

again assuming that $y \to 0$ as $x \to \infty$ and using (7.166). Further

$$\mathscr{F}_C\left\{\frac{d^2y}{dx^2}\right\} = -y'(0) - s^2\bar{y}_C(s), \tag{7.183}$$

with $y(x)$ and $y'(x) \to 0$ as $x \to \infty$.

(*c*) *Fourier sine transform* Here

$$\mathscr{F}_S\left\{\frac{dy}{dx}\right\} = \int_0^\infty \frac{dy}{dx}\sin(sx)\,dx \tag{7.184}$$

$$= [y(x)\sin(sx)]_0^\infty - s\int_0^\infty y(x)\cos(sx)\,dx \tag{7.185}$$

$$= -s\bar{y}_C(s) \tag{7.186}$$

and

$$\mathscr{F}_S\left\{\frac{d^2y}{dx^2}\right\} = sy'(0) - s^2\bar{y}_S(s), \tag{7.187}$$

provided $y(x)$ and $y'(x) \to 0$ as $x \to \infty$.

All the above transformations of derivatives may be extended to partial derivatives in the same way as in ***2*** of Section 7.7.

5. Differential equations

The application of Fourier transforms to the solution of partial differential equations is dealt with in the next chapter.

6. Integral equations

The integral equation

$$y(x) = f(x) + \int_{-\infty}^{\infty} y(u)g(x-u)\,du, \tag{7.188}$$

where f and g are given, is a type of Fredholm equation (the important point is that the limits are fixed whereas the Volterra equation of Section 7.11 has a variable upper limit of integration). Fourier transforming (7.188) and using the Convolution Theorem (7.172) gives

$$\bar{y}(s) = \bar{f}(s) + \bar{y}(s)\bar{g}(s), \tag{7.189}$$

whence

$$\bar{y}(s) = \frac{\bar{f}(s)}{1 - \bar{g}(s)}. \tag{7.190}$$

When $\bar{f}(s)$ and $\bar{g}(s)$ are known, $\bar{y}(s)$ may be inverted to give $y(x)$.

Example 14 Consider the integral equation

$$\int_{-\infty}^{\infty} f(x-u)f(u)\,du = \frac{1}{x^2+1}. \tag{7.191}$$

Taking the Fourier transform, we find

$$\mathscr{F}\left\{\int_{-\infty}^{\infty} f(x-u)f(u)\,du\right\} = \bar{f}(s)\bar{f}(s) = \mathscr{F}\left(\frac{1}{x^2+1}\right). \tag{7.192}$$

Now

$$\mathscr{F}\left(\frac{1}{x^2+1}\right) = \int_{-\infty}^{\infty} \frac{e^{-isx}}{x^2+1}\,dx \tag{7.193}$$

$$= 2\int_0^{\infty} \frac{\cos(sx)}{x^2+1}\,dx. \tag{7.194}$$

The integral in (7.170) and (7.171) is the derivative of that in (7.194) with k replacing s. Hence

$$\mathscr{F}\left(\frac{1}{x^2+1}\right) = 2\frac{\pi}{2}e^{-|s|}, \tag{7.195}$$

and from (7.192)

$$\bar{f}(s) = \sqrt{\pi}\,e^{-|s|/2}. \tag{7.196}$$

Inverting (7.196) we have, using (7.163),

$$f(x) = \frac{1}{2\pi}\int_{-\infty}^{\infty} \sqrt{\pi}\,e^{-|s|/2}e^{isx}\,ds \tag{7.197}$$

$$= \frac{1}{2\sqrt{\pi}}\left(\int_0^{\infty} e^{-s/2+isx}\,ds + \int_{-\infty}^{0} e^{s/2+isx}\,ds\right) \tag{7.198}$$

$$= \frac{1}{2\sqrt{\pi}}\left(\int_0^{\infty} e^{s(ix-\frac{1}{2})}\,ds + \int_0^{\infty} e^{-s(ix+\frac{1}{2})}\,ds\right) \tag{7.199}$$

$$= \frac{1}{2\sqrt{\pi}}\left(\frac{-1}{ix-\frac{1}{2}} + \frac{1}{ix+\frac{1}{2}}\right) \tag{7.200}$$

$$= \frac{1}{2\sqrt{\pi}\,(x^2+\frac{1}{4})}. \quad ◢ \tag{7.201}$$

Problems 7

1. Evaluate (i) $\mathscr{L}\{\cosh(ax)\cos(ax)\}$, (ii) $\mathscr{L}\{x^{\frac{3}{2}}\}$.
2. Use the Shift Theorem to obtain $\mathscr{L}\{e^x \sin^2 x\}$.
3. Find the Laplace inversions of the following functions:

$$\text{(i)}\ \frac{s}{(s^2+1)(s^2+4)}, \qquad \text{(ii)}\ \frac{1}{s(s-1)^3}.$$

4. Use the Laplace transform to evaluate

$$\text{(i)}\ \int_0^\infty e^{-x}\sin x\,dx, \quad \text{(ii)}\ \int_0^\infty e^{-x}x\sin x\,dx,$$

$$\text{(iii)}\ \int_0^\infty e^{-x}J_0(x)\,dx.$$

5. Show that

$$\mathscr{L}\{x^\alpha\} = \Gamma(\alpha+1)/s^{\alpha+1},$$

where $\alpha > -1$ is a real constant.

6. Use the Convolution Theorem to find the function $f(x)$ such that

$$\int_0^x \frac{f(u)}{\sqrt{(x-u)}}\,du = 1 + \sqrt{x},$$

for $x > 0$.

7. Use the Convolution Theorem to evaluate $\int_0^x J_0(u)J_0(x-u)\,du$.
8. Show that

$$\mathscr{L}\{xH(x-a)\} = \frac{e^{-sa}(1+sa)}{s^2},$$

for $a > 0$.

9. Solve by the Laplace transform method, where $D \equiv d/dx$:
   (i) $(D^2+3D+2)y = 4e^{-3x}$ with $y(0) = 1$, $y'(0) = 2$,
   (ii) $(D^2+4)y = H(x-2)$ with $y(0) = 1$, $y'(0) = 2$,
   (iii) $(D+1)y = \delta(x-a)$ with $y(0) = 1$, $a > 0$.
10. Use the Laplace transform to solve the equations

$$(D-2)x + Dy = 2,$$

$$(2D-1)x - (4D-3)y = 7$$

for $x(t)$ and $y(t)$, where $D \equiv d/dt$, with $x(0) = 0$, $y(0) = 1$.

11. Express the solution of

$$x\frac{d^2y}{dx^2} + (1-x)\frac{dy}{dx} + \lambda y = 0,$$

where λ is a real constant, in the form of an inverse Laplace transform, and hence find the solution when $\lambda = 2$.

12. Use the Bromwich contour method to find the inverse Laplace transforms of

$$\text{(i)}\ \frac{1}{(s+1)(s-2)^2}, \quad \text{(ii)}\ \frac{s}{(s+1)^3(s-1)^2}.$$

13. Find, using the Convolution Theorem, the solution of the integral equation

$$y(x) = \sin(3x) + \int_0^x \sin(x-u)y(u)\,du.$$

14. Using the result

$$\int_0^\infty e^{-\lambda x^2} \cos(sx)\,dx = \surd(\pi/4\lambda)e^{-s^2/4\lambda},$$

find the Fourier transform of $e^{-\lambda x^2}$. Hence derive the Fourier transform of the delta-function $\delta(x)$ by a limiting process using the definition

$$\delta(x) = \lim_{n\to\infty} \surd(n/\pi)e^{-nx^2}.$$

15. Show that the Fourier transform of $e^{-a|x|}$ $(a > 0)$ is $2a/(s^2 + a^2)$ and hence derive the transform of $x^2 e^{-a|x|}$.
16. Find the Fourier sine and cosine transforms of (i) e^{-2x}, (ii) $1/x$.
17. Using the first result of Problem 15 with $a = 1$, show from the Inversion Theorem that

$$\int_{-\infty}^\infty \frac{\cos x}{1+x^2}\,dx = \frac{\pi}{e}.$$

18. Using the Fourier Convolution Theorem and the first result of Problem 15, solve the integral equation

$$\int_{-\infty}^\infty \frac{f(u)}{(x-u)^2+1}\,du = \frac{1}{x^2+4}.$$

8
Partial differential equations

8.1 Introduction

Partial differential equations occur frequently in the formulation of basic laws of nature and in the mathematical study of a wide variety of problems. In general, since we live in a universe of three space dimensions and one time dimension, the dependent variable, u, say, is a function of at most four independent variables. Relative to a cartesian coordinate system, therefore, we have

$$u = u(x, y, z, t). \tag{8.1}$$

The relationship between u and its partial derivatives $u_x = \partial u/\partial x$, $u_{xy} = \partial^2 u/\partial x\,\partial y, \ldots,$

$$f(x, y, z, t; u, u_x, u_y, u_{xx}, u_{xy}, \ldots) = 0, \tag{8.2}$$

where f is some function, is called a partial differential equation. If f is linear in each of the variables $u, u_x, u_y, u_{xx}, \ldots$ and if the coefficients of each of these variables are functions only of the *independent variables* x, y, z, t, then the equation (8.2) is said to be linear. Equations which are not of this type are said to be non-linear. In general, the order of the partial differential equation is defined by the order of the highest-order partial derivative in the equation.

8.2 Principle of Superposition

As with linear ordinary differential equations (see Section 3.1), the Principle of Superposition applies also to linear partial differential equations. As an illustration, consider the one-dimensional wave equation

$$\frac{\partial^2 u}{\partial x^2} = \frac{1}{c^2}\frac{\partial^2 u}{\partial t^2}, \tag{8.3}$$

where c is a constant. It is easily verified that the two solutions of (8.3) are

$$u_1 = f(x + ct), \quad u_2 = g(x - ct), \tag{8.4}$$

where f and g are arbitrary functions, at least twice differentiable, of the arguments $(x + ct)$ and $(x - ct)$, respectively. Accordingly, by the Principle of Superposition,

$$u = f(x + ct) + g(x - ct) \tag{8.5}$$

is also a solution of (8.3). We shall show in Section 8.6 that (8.5) is, in fact, the general solution of (8.3).

The Principle of Superposition, however, does not apply to non-linear equations. There is no general method of obtaining analytical solutions of non-linear partial differential equations and numerical procedures are usually required for their solution. Because of the difficulty in solving such equations, approximations are sometimes made which linearise the equations. Provided the physical implications of such approximations are understood, the analytical solutions of the linearised equation (which are much easier to obtain) often give a valuable insight into the solution of the original problem.

8.3 Some important equations

Many physical processes are described to some degree of accuracy by linear second-order partial differential equations and it is principally the analytical solution of this type of equation with which we shall be concerned in this chapter. Some of these equations are listed here, making use of the conventional notation that ∇^2 represents the Laplacian operator which in three-dimensional cartesian coordinates takes the form

$$\nabla^2 = \frac{\partial^2}{\partial x^2} + \frac{\partial^2}{\partial y^2} + \frac{\partial^2}{\partial z^2}. \tag{8.6}$$

The form of ∇^2 in other coordinate systems has been discussed in Chapter 1.

(a) The wave equation

$$\nabla^2 u = \frac{1}{c^2}\frac{\partial^2 u}{\partial t^2}, \tag{8.7}$$

where c is a constant.

(b) The heat conduction, or diffusion, equation

$$\nabla^2 u = \frac{1}{K}\frac{\partial u}{\partial t}, \tag{8.8}$$

where K is a constant.

(c) Laplace's equation

$$\nabla^2 u = 0. \tag{8.9}$$

(d) Helmholtz's equation

$$\nabla^2 u + \lambda u = 0, \tag{8.10}$$

where λ is a constant.

(e) Poisson's equation

$$\nabla^2 u = f(x, y, z). \tag{8.11}$$

(f) Schrödinger's equation

$$\nabla^2 u + \alpha[E - V(x, y, z)]u = 0, \tag{8.12}$$

where α and E are constants and V is the potential function.

(g) The Klein–Gordon equation

$$\Box u + \lambda u = 0, \tag{8.13}$$

where the operator $\Box$, the d'Alembertian, is defined by

$$\Box = \nabla^2 - \frac{1}{c^2}\frac{\partial^2}{\partial t^2}, \tag{8.14}$$

and λ is a constant.

Higher-order equations also arise. For example, the biharmonic wave equation

$$\nabla^4 u = \nabla^2(\nabla^2 u) = -\frac{1}{K}\frac{\partial^2 u}{\partial t^2}, \tag{8.15}$$

where K is a constant, describes the transverse oscillations of a thin plate.

First-order equations of the general form

$$P(x, y)\frac{\partial u}{\partial x} + Q(x, y)\frac{\partial u}{\partial y} = R(x, y)u^n \tag{8.16}$$

also occur in a variety of physical problems (for example, in fluid flow, kinetic theory of gases, transport processes). We note that (8.16) is linear in u if $n = 0$ or 1, but non-linear otherwise.

8.4 Linear second-order equations in two independent variables

In the previous section, a number of physically important equations were listed. Many of these equations, when only two independent variables x and y are present, are special cases of the general linear second-order equation

$$a\frac{\partial^2 u}{\partial x^2}+2h\frac{\partial^2 u}{\partial x\,\partial y}+b\frac{\partial^2 u}{\partial y^2}+2f\frac{\partial u}{\partial x}+2g\frac{\partial u}{\partial y}+cu=F(x, y). \quad (8.17)$$

The coefficients of u and its derivatives, namely a, h, b, f, g and c, are constants or given functions of x and y, and $F(x, y)$ is a given function. The physical interpretation of the independent variables depends on the particular situation. For example, Laplace's equation (8.9) in two space dimensions, x and y, is a special case of (8.17) obtained by putting $a=1$, $h=0$, $b=1$, $f=g=c=0$, $F(x, y)=0$. On the other hand, the heat conduction equation (8.8) in one space dimension, x, is obtained from (8.17) by associating the variable y in (8.17) with time t and then taking $a=1$, $h=0$, $b=f=c=0$, $g=-1/2K$, $F(x, t)=0$.

We note that the homogeneous form of (8.17) (obtained by putting $F(x, y)=0$) resembles the equation of a general conic, that is,

$$ax^2+2hxy+by^2+2fx+2gy+c=0. \quad (8.18)$$

Equation (8.18) represents an ellipse (the circle being a special case), parabola or hyperbola if $ab-h^2$ is positive, zero or negative, respectively. Hence we term (8.17) as being of

$$\text{elliptic type} \quad \text{if} \quad ab-h^2>0, \quad (8.19)$$

$$\text{parabolic type} \quad \text{if} \quad ab-h^2=0, \quad (8.20)$$

or

$$\text{hyperbolic type} \quad \text{if} \quad ab-h^2<0. \quad (8.21)$$

Clearly Laplace's equation (8.9) for which $a=1$, $h=0$, and $b=1$ (as above) has $ab-h^2=1>0$ and is therefore of elliptic type. Similarly, the heat conduction equation (8.8) in x and t for which (as we have seen) $a=1$, $h=0$, and $b=0$ has $ab-h^2=0$ and is therefore of parabolic type. In a similar way, it is found that the wave equation (8.7) in x and t is of hyperbolic type.

When a, b and h are functions of x and y, the nature of the equation may change from one region of the (x, y) plane to another. For example, the equation $y\,\partial^2 u/\partial x^2+2x\,\partial^2 u/\partial x\,\partial y+x\,\partial^2 u/\partial y^2=0$ for which $ab-h^2=xy-x^2=x(y-x)$, is elliptic if $xy-x^2>0$ (that is, if

$x > 0$ and $y - x > 0$, or if $x < 0$ and $y - x < 0$), and is hyperbolic if $xy - x^2 < 0$ (that is, if $x > 0$ and $y - x < 0$, or if $x < 0$ and $y - x > 0$).

8.5 Boundary conditions

The general solutions of partial differential equations involve arbitrary functions (see (8.5)), whereas the general solutions of ordinary differential equations involve arbitrary constants. This shows that there is an infinity of possible solutions and only by specifying the boundary conditions can we obtain a specific solution. For this to represent a physical situation, we require the problem as defined by the partial differential equation and its boundary conditions to be well posed. This criterion is satisfied if (*a*) the solution is unique, and (*b*) small changes in the boundary conditions and the various coefficients in the equation give rise to only small changes in the solution. When (*b*) holds the solution is said to be stable.

Much work has been carried out to determine the various types of boundary conditions which, when imposed on a partial differential equation, lead to a unique stable solution. In general, there are three main types of boundary conditions which arise:

(*A*) *Dirichlet conditions* These specify the dependent variable u at each point of the boundary of the region within which a solution is required. For example, in two dimensions, u is specified at every point on a boundary curve C of a plane region R.

The boundary value problem requiring the solution of $\nabla^2 u = 0$ (in the appropriate number of dimensions) within a region R, subject to u being given on the boundary of R, is called the Dirichlet problem.

(*B*) *Neumann conditions* In this case, the values of the normal derivative $\partial u/\partial n$ on the boundary of R are specified.

(*C*) *Cauchy conditions* These arise mainly in time-developing situations where u and $\partial u/\partial t$ are specified along the line $t = 0$. Such conditions, therefore, are called initial conditions.

Cauchy conditions are generally met in conjunction with hyperbolic equations, whereas Dirichlet and Neumann conditions relate specifically to elliptic and parabolic equations. Applying the wrong type of boundary conditions to a particular equation can lead to an over-determined system or an under-determined system. We shall not attempt to discuss these problems here since we are more concerned with the techniques of solution assuming that the given boundary conditions are of the appropriate type for a well-posed problem.

8.6 Method of characteristics

1. First-order equations

The method of characteristics can be applied usefully to both linear and non-linear equations. To illustrate the technique, we take the simplest form of (8.16) by letting $n = 0$, so that

$$P(x, y)\frac{\partial u}{\partial x} + Q(x, y)\frac{\partial u}{\partial y} = R(x, y). \tag{8.22}$$

Consider now the family of curves defined by the equation

$$\frac{dx}{P} = \frac{dy}{Q} = dt \text{ (say)}, \tag{8.23}$$

where we write $P = P(x, y)$ and $Q = Q(x, y)$. Then considering the infinitesimal triangle (Figure 8.1), where ds is the infinitesimal element of arc length of a typical curve C of the family given by

$$ds^2 = dx^2 + dy^2, \tag{8.24}$$

we have, using (8.23),

$$ds^2 = \frac{P^2}{Q^2}dy^2 + dy^2 \tag{8.25}$$

$$= dx^2 + \frac{Q^2}{P^2}dx^2. \tag{8.26}$$

Hence

$$\frac{dx}{P} = \frac{dy}{Q} = \frac{ds}{\sqrt{(P^2 + Q^2)}} = dt. \tag{8.27}$$

Now from (8.23), the curves are defined by

$$\frac{dy}{dx} = \frac{Q}{P}. \tag{8.28}$$

Figure 8.1

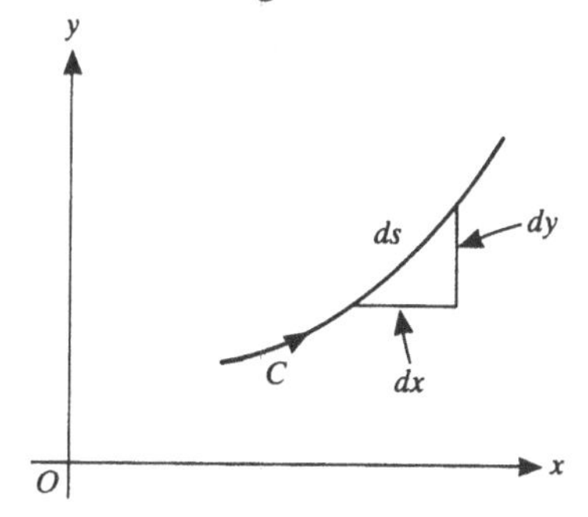

Multiplying (8.22) by ds and using (8.27), we have

$$P\,ds\frac{\partial u}{\partial x}+Q\,ds\frac{\partial u}{\partial y}=R\,ds \tag{8.29}$$

and hence

$$\sqrt{(P^2+Q^2)}\,dx\frac{\partial u}{\partial x}+\sqrt{(P^2+Q^2)}\,dy\frac{\partial u}{\partial y}=R\,ds. \tag{8.30}$$

The total differential of $u(x, y)$ is

$$du=\frac{\partial u}{\partial x}dx+\frac{\partial u}{\partial y}dy \tag{8.31}$$

and so (8.30) takes the form

$$du=\frac{R}{\sqrt{(P^2+Q^2)}}ds. \tag{8.32}$$

Using (8.27) this becomes

$$du=R\,dt. \tag{8.33}$$

Finally then (8.23) may be written as

$$\frac{dx}{P}=\frac{dy}{Q}=\frac{du}{R}(=dt), \tag{8.34}$$

from which u can be found by integration. This integration is clearly along any curve C defined by (8.23). These curves form a special family relating to the given differential equation and are called the characteristic curves (or, simply, the characteristics). The method of characteristics requires that the region over which (8.22) is to be solved should be filled with characteristic curves. The following examples illustrate the method in some simple cases.

Example 1 To solve

$$x\frac{\partial u}{\partial x}+y\frac{\partial u}{\partial y}=2xy, \tag{8.35}$$

given $u=2$ on $y=x^2$.

The basic equation (8.34) gives (since $P=x$, $Q=y$, $R=2xy$)

$$\frac{dx}{x}=\frac{dy}{y}=\frac{du}{2xy} \tag{8.36}$$

and the characteristics are defined by

$$dy/dx=y/x \tag{8.37}$$

(see (8.28)). The solution of (8.37) is the family of curves

$$y/x = K, \tag{8.38}$$

where K is a parameter. Hence the characteristics are straight lines through the origin with arbitrary gradients. From (8.36) we also have on any characteristic curve

$$du/2xy = dy/y \tag{8.39}$$

whence, cancelling y and using (8.38) to obtain $dy = K\,dx$, we have

$$du = 2xK\,dx. \tag{8.40}$$

Integrating we find

$$u = Kx^2 + A, \tag{8.41}$$

where A is an arbitrary constant of integration. This constant is, in general, different on each characteristic curve. Further, each characteristic curve has a different value of K. Hence, as K varies, A varies and we may write $A = f(K)$, where f is an arbitrary function to be determined. Writing $A = f(K)$ in (8.41) and eliminating K using (8.38), we see that the general solution of (8.35) is

$$u = xy + f(y/x). \tag{8.42}$$

Since on the curve $y = x^2$, $u = 2$ (given), we have

$$2 = x^3 + f(x). \tag{8.43}$$

Hence

$$f(x) = 2 - x^3 \tag{8.44}$$

giving

$$f(y/x) = 2 - (y/x)^3. \tag{8.45}$$

Substituting (8.45) into (8.42), we see that the required solution of (8.35) is

$$u = xy + 2 - (y/x)^3. \quad ◢ \tag{8.46}$$

Example 2 Consider now the homogeneous equation

$$2xy\frac{\partial u}{\partial x} + (x^2 + y^2)\frac{\partial u}{\partial y} = 0. \tag{8.47}$$

Equation (8.34) becomes (since $R = 0$)

$$\frac{dx}{2xy} = \frac{dy}{(x^2 + y^2)} = \frac{du}{0}. \tag{8.48}$$

Hence the characteristics are defined by

$$\frac{dy}{dx} = \frac{(x^2 + y^2)}{2xy}, \tag{8.49}$$

which on putting $y = vx$ gives

$$x\frac{dv}{dx} = \frac{(1 - v^2)}{2v}. \tag{8.50}$$

Separating the variables and integrating, we find

$$1 - v^2 = K/x, \tag{8.51}$$

where K is a parameter, and hence

$$\frac{x^2 - y^2}{x} = K. \tag{8.52}$$

Also (8.48) requires, for finiteness, that

$$du = 0 \tag{8.53}$$

or

$$u = \text{constant} = f(K), \tag{8.54}$$

as explained in Example 1. Hence, using (8.52), we find that the general solution is

$$u = f\left(\frac{x^2 - y^2}{x}\right). \tag{8.55}$$

Suppose now that some boundary condition is given, say

$$u = e^{x/(x-y)} \quad \text{when} \quad x + y = 1. \tag{8.56}$$

Then

$$e^{x/(2x-1)} = f\left(\frac{2x - 1}{x}\right). \tag{8.57}$$

Letting $(2x - 1)/x = w$, say, we have

$$f(w) = e^{1/w}, \tag{8.58}$$

which defines the function f. Consequently (8.55) subject to (8.56) becomes, using the form of f found in (8.58),

$$u = e^{x/(x^2-y^2)}. \quad ◢ \tag{8.59}$$

At the beginning of this section, we considered the simple form of (8.16) with $n = 0$. Equations of the general type (8.16) can easily be reduced to the form (8.22) by letting $u^{1-n} = v$, for $n \neq 1$, so that (8.16) becomes

$$P\frac{\partial v}{\partial x} + Q\frac{\partial v}{\partial y} = (1-n)R. \tag{8.60}$$

This equation can be solved by the method described in this subsection. If $n = 1$, then $v = \ln u$ will reduce equation (8.16) to the required form of (8.22). More generally, the method is applicable to equations where R is a function of x, y and u. We illustrate this by an example.

Example 3 To solve

$$x\frac{\partial u}{\partial x} + y\frac{\partial u}{\partial y} = x\mathrm{e}^{-u}, \tag{8.61}$$

with $u = 0$ on $y = x^2$.

Following (8.34) with $R = x\mathrm{e}^{-u}$, we have

$$\frac{dx}{x} = \frac{dy}{y} = \frac{du}{x\mathrm{e}^{-u}}. \tag{8.62}$$

As in Example 1, the characteristics are

$$y/x = K. \tag{8.63}$$

We also have from (8.62)

$$dx = \mathrm{e}^{u}\, du, \tag{8.64}$$

which on integration gives

$$\mathrm{e}^{u} = x + f(K). \tag{8.65}$$

Hence using (8.63),

$$\mathrm{e}^{u} = x + f(y/x). \tag{8.66}$$

Now $u = 0$ when $y = x^2$, giving

$$f(x) = 1 - x. \tag{8.67}$$

From (8.66), therefore, we have

$$\mathrm{e}^{u} = x + 1 - y/x, \tag{8.68}$$

whence

$$u = \ln(x + 1 - y/x). \tag{8.69}$$

2. Second-order equations

One of the simplest applications of characteristics to second-order equations occurs when the equation may be factorised into two first-order equations. We illustrate this by the following example.

Example 4 To solve

$$\frac{\partial^2 u}{\partial x^2}+2x\frac{\partial^2 u}{\partial x\,\partial y}+x^2\frac{\partial^2 u}{\partial y^2}+\frac{\partial u}{\partial y}=0, \tag{8.70}$$

given that $\partial u/\partial x = y^2$ on $x=0$ and $u=\mathrm{e}^y$ on $x=0$.

Consider the operator

$$L=\frac{\partial}{\partial x}+x\frac{\partial}{\partial y}. \tag{8.71}$$

Then

$$\left(\frac{\partial}{\partial x}+x\frac{\partial}{\partial y}\right)\left(\frac{\partial}{\partial x}+x\frac{\partial}{\partial y}\right)u=\frac{\partial^2 u}{\partial x^2}+2x\frac{\partial^2 u}{\partial x\,\partial y}+x^2\frac{\partial^2 u}{\partial y^2}+\frac{\partial u}{\partial y}, \tag{8.72}$$

and hence (8.70) may be written as

$$L^2u=0. \tag{8.73}$$

Now writing

$$Lu=w, \tag{8.74}$$

(8.73) becomes

$$L^2u=L(Lu)=Lw=0, \tag{8.75}$$

or, using (8.71),

$$\frac{\partial w}{\partial x}+x\frac{\partial w}{\partial y}=0. \tag{8.76}$$

The characteristics of (8.76) are found from

$$\frac{dx}{1}=\frac{dy}{x}=\frac{dw}{0}, \tag{8.77}$$

whence

$$y=\tfrac{1}{2}x^2+K. \tag{8.78}$$

Also $dw=0$ from (8.77), which implies

$$w=f(K), \tag{8.79}$$

where f is an arbitrary function. The general solution of (8.76) is therefore

$$w=f(y-\tfrac{1}{2}x^2). \tag{8.80}$$

Since $\partial u/\partial x = y^2$ on $x = 0$, we have, from (8.74),

$$w = \frac{\partial u}{\partial x} + x\frac{\partial u}{\partial y} = \frac{\partial u}{\partial x} = y^2 \quad \text{on} \quad x = 0. \tag{8.81}$$

Using this in (8.80) gives

$$y^2 = f(y) \tag{8.82}$$

and hence, again from (8.80), we find

$$w = (y - \tfrac{1}{2}x^2)^2. \tag{8.83}$$

Having found w, we can now find u from (8.74) using the method of characteristics. Writing out (8.74), using (8.71) and (8.83), we have

$$\frac{\partial u}{\partial x} + x\frac{\partial u}{\partial y} = (y - \tfrac{1}{2}x^2)^2, \tag{8.84}$$

whence

$$\frac{dx}{1} = \frac{dy}{x} = \frac{du}{(y - \frac{1}{2}x^2)^2}. \tag{8.85}$$

The characteristics are therefore

$$y = \tfrac{1}{2}x^2 + \bar{K}, \tag{8.86}$$

where $\bar{K}$ is a new constant of integration. Also

$$du = (y - \tfrac{1}{2}x^2)^2\, dx \tag{8.87}$$

$$= (\tfrac{1}{2}x^2 + \bar{K} - \tfrac{1}{2}x^2)^2\, dx = \bar{K}^2\, dx, \tag{8.88}$$

whence

$$u = \bar{K}^2 x + g(\bar{K}), \tag{8.89}$$

where g is an arbitrary function. Substituting for $\bar{K}$ using (8.86) we have the general solution

$$u = x(y - \tfrac{1}{2}x^2)^2 + g(y - \tfrac{1}{2}x^2). \tag{8.90}$$

Now $u = e^y$ on $x = 0$ (given) and hence from (8.90)

$$e^y = g(y). \tag{8.91}$$

Substituting for $g(y)$ in (8.90), we have finally

$$u = x(y - \tfrac{1}{2}x^2)^2 + e^{(y - \frac{1}{2}x^2)}. \tag{8.92}$$

We now discuss the method of characteristics as applied to second-order constant coefficient equations which are not factorisable into two first-order equations as in Example 4. For this purpose we return to the general linear form (8.17) and specialise to $f = g = c = 0$ and $F = 0$ to obtain

$$a\frac{\partial^2 u}{\partial x^2} + 2h\frac{\partial^2 u}{\partial x\,\partial y} + b\frac{\partial^2 u}{\partial y^2} = 0. \tag{8.93}$$

This equation, known as Euler's equation, is of one of the types described by (8.19)–(8.21). Here we take a, h and b to be *constants.*

Defining new independent variables

$$\xi = px + qy, \quad \eta = rx + sy, \tag{8.94}$$

where p, q, r and s are constants, we have

$$\frac{\partial u}{\partial x} = \frac{\partial u}{\partial \xi}\frac{\partial \xi}{\partial x} + \frac{\partial u}{\partial \eta}\frac{\partial \eta}{\partial x} = p\frac{\partial u}{\partial \xi} + r\frac{\partial u}{\partial \eta}, \tag{8.95}$$

$$\frac{\partial u}{\partial y} = \frac{\partial u}{\partial \xi}\frac{\partial \xi}{\partial y} + \frac{\partial u}{\partial \eta}\frac{\partial \eta}{\partial y} = q\frac{\partial u}{\partial \xi} + s\frac{\partial u}{\partial \eta}. \tag{8.96}$$

Hence

$$\frac{\partial^2 u}{\partial x^2} = \frac{\partial}{\partial x}\left(\frac{\partial u}{\partial x}\right) = \left(p\frac{\partial}{\partial \xi} + r\frac{\partial}{\partial \eta}\right)\left(p\frac{\partial u}{\partial \xi} + r\frac{\partial u}{\partial \eta}\right) \tag{8.97}$$

$$= p^2\frac{\partial^2 u}{\partial \xi^2} + 2pr\frac{\partial^2 u}{\partial \xi\,\partial \eta} + r^2\frac{\partial^2 u}{\partial \eta^2}, \tag{8.98}$$

$$\frac{\partial^2 u}{\partial y^2} = \frac{\partial}{\partial y}\left(\frac{\partial u}{\partial y}\right) = \left(q\frac{\partial}{\partial \xi} + s\frac{\partial}{\partial \eta}\right)\left(q\frac{\partial u}{\partial \xi} + s\frac{\partial u}{\partial \eta}\right) \tag{8.99}$$

$$= q^2\frac{\partial^2 u}{\partial \xi^2} + 2qs\frac{\partial^2 u}{\partial \xi\,\partial \eta} + s^2\frac{\partial^2 u}{\partial \eta^2} \tag{8.100}$$

and

$$\frac{\partial^2 u}{\partial x\,\partial y} = \frac{\partial}{\partial x}\left(\frac{\partial u}{\partial y}\right) = \left(p\frac{\partial}{\partial \xi} + r\frac{\partial}{\partial \eta}\right)\left(q\frac{\partial u}{\partial \xi} + s\frac{\partial u}{\partial \eta}\right) \tag{8.101}$$

$$= pq\frac{\partial^2 u}{\partial \xi^2} + (rq + sp)\frac{\partial^2 u}{\partial \xi\,\partial \eta} + rs\frac{\partial^2 u}{\partial \eta^2}. \tag{8.102}$$

Using (8.98), (8.100) and (8.102), (8.93) becomes

$$(ap^2 + 2hpq + bq^2)\frac{\partial^2 u}{\partial \xi^2} + 2[apr + bsq + h(rq + sp)]\frac{\partial^2 u}{\partial \xi\,\partial \eta}$$

$$+ (ar^2 + 2hrs + bs^2)\frac{\partial^2 u}{\partial \eta^2} = 0. \tag{8.103}$$

We may now choose the constants p, q, r and s to simplify the equation. Let $p = r = 1$ and choose q and s to be the two roots λ_1 and λ_2, say, of the quadratic equation

$$a + 2h\lambda + b\lambda^2 = 0. \tag{8.104}$$

Then (8.103) becomes

$$2[a + h(\lambda_1 + \lambda_2) + b\lambda_1\lambda_2]\frac{\partial^2 u}{\partial\xi\,\partial\eta} = 0. \tag{8.105}$$

From (8.104),

$$\lambda_1 + \lambda_2 = -2h/b, \tag{8.106}$$

$$\lambda_1\lambda_2 = a/b. \tag{8.107}$$

Hence (8.105) finally has the form

$$\frac{4}{b}(ab - h^2)\frac{\partial^2 u}{\partial\xi\,\partial\eta} = 0. \tag{8.108}$$

Provided now $b \neq 0$ and the equation is not parabolic in nature ($ab - h^2 \neq 0$), (8.108) becomes

$$\partial^2 u/\partial\xi\,\partial\eta = 0, \tag{8.109}$$

where, making use of our choices of p, q, r and s,

$$\xi = x + \lambda_1 y, \quad \eta = x + \lambda_2 y. \tag{8.110}$$

Now (8.109) may be integrated immediately, first with respect to, say, ξ giving $\partial u/\partial\eta = m(\eta)$, where $m(\eta)$ is an arbitrary function, and then with respect to η giving

$$u = f(\xi) + g(\eta), \tag{8.111}$$

where f and g are arbitrary functions. Using (8.110), we have the general solution

$$u = f(x + \lambda_1 y) + g(x + \lambda_2 y). \tag{8.112}$$

The curves of constant ξ and constant η are called the characteristic curves. Hence

$$x + \lambda_1 y = \text{constant}, \quad x + \lambda_2 y = \text{constant} \tag{8.113}$$

are the characteristics of the Euler equation (8.93) with a, b and h as constants, and the form (8.109) is said to be the *canonic* form of the original equation. We see that by choosing particular curves (the characteristics defined by (8.113)) the equation can be brought into a simpler form which may be integrated more easily.

The same type of analysis can be applied when the equation is parabolic ($ab - h^2 = 0$), but a different choice of p, q, r and s is needed to simplify the equation. We shall not discuss this here.

Another feature illustrated by this analysis is that the basic equation (8.104) defining λ clearly has real roots if $ab - h^2 < 0$ and complex roots if $ab - h^2 > 0$. Hence we see that hyperbolic equations have real characteristics, whereas elliptic equations have no real characteristic curves.

We conclude this section with two examples.

Example 5 The wave equation

$$\frac{\partial^2 u}{\partial x^2} - \frac{1}{c^2}\frac{\partial^2 u}{\partial t^2} = 0 \tag{8.114}$$

is a special case of (8.17) with $a = 1$, $h = 0$ and $b = -1/c^2$. Hence $ab - h^2 = -1/c^2 < 0$ and the equation is therefore hyperbolic (as mentioned in Section 8.4). Accordingly (8.104) becomes

$$1 - \lambda^2/c^2 = 0, \tag{8.115}$$

giving

$$\lambda_1 = c, \quad \lambda_2 = -c. \tag{8.116}$$

The characteristics are therefore

$$\xi = x + ct = \text{constant}, \tag{8.117}$$

$$\eta = x - ct = \text{constant}, \tag{8.118}$$

and the canonic form is

$$\frac{\partial^2 u}{\partial \xi\, \partial \eta} = 0. \tag{8.119}$$

Using (8.112) we have the general solution

$$u = f(x + ct) + g(x - ct), \tag{8.120}$$

where f and g are arbitrary functions, as in (8.5). ◢

Example 6 Consider the equation

$$2\frac{\partial^2 u}{\partial x^2} - 3\frac{\partial^2 u}{\partial x\, \partial y} + \frac{\partial^2 u}{\partial y^2} = y. \tag{8.121}$$

Then $ab - h^2 = 2 - (-\frac{3}{2})^2 = -\frac{1}{4} < 0$ and so (8.121) is hyperbolic. The equation giving the values of λ, (8.104), is

$$2 - 3\lambda + \lambda^2 = 0, \tag{8.122}$$

whence

$$\lambda = 1 \quad \text{or} \quad \lambda = 2. \tag{8.123}$$

The characteristic curves are

$$\xi = x + 2y = \text{constant}, \tag{8.124}$$

$$\eta = x + y = \text{constant}. \tag{8.125}$$

The equation reduces, by (8.108), to the canonic form

$$\frac{4}{1}\left(-\frac{1}{4}\right)\frac{\partial^2 u}{\partial \xi\, \partial \eta} = y = \xi - \eta, \tag{8.126}$$

or

$$\partial^2 u / \partial \xi\, \partial \eta = \eta - \xi. \tag{8.127}$$

This equation may be readily integrated, first with respect to η giving

$$\partial u / \partial \xi = \tfrac{1}{2}\eta^2 - \xi\eta + f(\xi), \tag{8.128}$$

where f is an arbitrary function, and secondly with respect to ξ giving

$$u = \tfrac{1}{2}\eta^2\xi - \tfrac{1}{2}\xi^2\eta + F(\xi) + G(\eta), \tag{8.129}$$

where F and G are arbitrary functions. Expressed in terms of x and y, using (8.124) and (8.125), we have

$$u = -\tfrac{1}{2}y(x + y)(x + 2y) + F(x + 2y) + G(x + y). \;◢ \tag{8.130}$$

8.7 Separation of variables

This method applies only to linear equations and makes use of the Principle of Superposition (see Section 8.2) in building up a linear combination of individual solutions to form a solution satisfying the boundary conditions. The basic approach in attempting to solve equations (in, say, two independent variables x and y) in this way is to write the dependent variable $u(x, y)$ in the separable form

$$u(x, y) = X(x)Y(y), \tag{8.131}$$

where X and Y are functions only of x and of y respectively. In many cases the partial differential equation reduces to two ordinary differential equations for X and Y. Similar considerations apply to equations in three or more independent variables. The following examples illustrate the method both for first-order and second-order equations.

Example 7 To solve

$$\frac{\partial u}{\partial x} - x^2 \frac{\partial u}{\partial y} = 2x^2 y u, \tag{8.132}$$

given $u(x, 0) = \cosh x^3$.

Now writing

$$u(x, y) = X(x)Y(y), \tag{8.133}$$

equation (8.132) becomes

$$X'Y - x^2 XY' = 2x^2 yXY, \tag{8.134}$$

where the primes denote ordinary differentiation with respect to the appropriate variable ($X' = dX/dx$, $Y' = dY/dy$). Equation (8.134) may be written as

$$\frac{1}{x^2}\frac{X'}{X} = 2y + \frac{Y'}{Y}, \tag{8.135}$$

from which we see that the left-hand side is a function of x only, whilst the right-hand side is a function of y only. Hence the equation has been separated with terms dependent only on x on one side and only on y on the other. For (8.135) to be satisfied, each side must be equal to a constant (α, say) so that

$$\frac{1}{x^2}\frac{X'}{X} = \alpha, \quad \frac{Y'}{Y} + 2y = \alpha. \tag{8.136}$$

This is a pair of ordinary differential equations for the dependent variables $X(x)$ and $Y(y)$. Integrating the first of the equations in (8.136), we find

$$X(x) = A\mathrm{e}^{\alpha x^3/3}, \tag{8.137}$$

while integrating the second, we have

$$Y(y) = B\mathrm{e}^{\alpha y - y^2}, \tag{8.138}$$

where A and B are arbitrary constants. Hence, from (8.133),

$$u(x, y) = X(x)Y(y) = AB\mathrm{e}^{-y^2}\mathrm{e}^{\alpha(x^3/3+y)} \tag{8.139}$$

$$= C\mathrm{e}^{-y^2}\mathrm{e}^{\alpha(x^3/3+y)}, \tag{8.140}$$

where C is an arbitrary constant. We can have an infinity of such solutions with different choices of C and α. By the Principle of Superposition, the linear combination

$$u = \mathrm{e}^{-y^2} \sum_{\substack{\text{all } C \\ \text{all } \alpha}} C\mathrm{e}^{\alpha(x^3/3+y)} \tag{8.141}$$

is the general solution of (8.132). The values of C and α in (8.141) are arbitrary at this stage and must be found by applying the boundary condition. It is more convenient, however, to write (8.141) as

$$u = e^{-y^2} F(x^3/3 + y), \tag{8.142}$$

since the sum is some function F of $x^3/3 + y$.

Now $u(x, 0) = \cosh x^3$ so that, from (8.142),

$$\cosh x^3 = F(x^3/3). \tag{8.143}$$

Letting $x^3/3 = \theta$, then

$$F(\theta) = \cosh(3\theta). \tag{8.144}$$

Hence

$$F(x^3/3 + y) = \cosh(x^3 + 3y) \tag{8.145}$$

and the required solution is, by (8.142),

$$u = e^{-y^2} \cosh(x^3 + 3y). \ ◢ \tag{8.146}$$

The separation of a partial differential equation into ordinary differential equations, as in (8.136), is not always possible, as the following example shows.

Example 8 To show that the equation

$$\frac{\partial^2 u}{\partial x^2} + \frac{\partial^2 u}{\partial y^2} + (x^2 + y^2)^2 u = 0 \tag{8.147}$$

is not separable in x and y but is separable in plane polar coordinates (r, θ).

Writing $u = X(x)Y(y)$, (8.147) becomes

$$\frac{X''}{X} + \frac{Y''}{Y} + x^4 + 2x^2y^2 + y^4 = 0 \tag{8.148}$$

or

$$\left(\frac{X''}{X} + x^4\right) + 2x^2y^2 + \left(\frac{Y''}{Y} + y^4\right) = 0. \tag{8.149}$$

The first bracket is a function of x only and the second bracket a function of y only, but the product $2x^2y^2$ contains both x and y. Hence the equation is not separable.

From (1.102), the form of $\nabla^2 u$ in plane polar coordinates (r, θ) is

$$\nabla^2 u = \frac{1}{r}\frac{\partial}{\partial r}\left(r\frac{\partial u}{\partial r}\right) + \frac{1}{r^2}\frac{\partial^2 u}{\partial \theta^2}. \tag{8.150}$$

Hence (8.147) becomes

$$\frac{1}{r}\frac{\partial}{\partial r}\left(r\frac{\partial u}{\partial r}\right)+\frac{1}{r^2}\frac{\partial^2 u}{\partial \theta^2}+r^4 u=0. \tag{8.151}$$

Writing $u=R(r)\Theta(\theta)$ we have

$$\left[\frac{r}{R}\frac{d}{dr}\left(r\frac{dR}{dr}\right)+r^6\right]+\frac{1}{\Theta}\frac{d^2\Theta}{d\theta^2}=0. \tag{8.152}$$

The terms in square brackets are functions of r only and the remaining term is a function of θ only. Hence (8.152) is separable into two ordinary differential equations for $R(r)$ and $\Theta(\theta)$,

$$\frac{r}{R}\frac{d}{dr}\left(r\frac{dR}{dr}\right)+r^6=\alpha, \quad \frac{1}{\Theta}\frac{d^2\Theta}{d\theta^2}=-\alpha, \tag{8.153}$$

where α is called the separation constant. ◢

Example 9 The temperature distribution $T(r, t)$ in a homogeneous sphere of radius a satisfies the equation

$$\nabla^2 T=\frac{1}{K}\frac{\partial T}{\partial t}, \tag{8.154}$$

for $0 \leqslant r \leqslant a$, $t>0$, where K is a positive constant and ∇^2 is in spherical polar coordinates (r, θ, ϕ). Show that the substitution $u(r, t)=rT(r, t)$ transforms this equation into

$$\frac{\partial^2 u}{\partial r^2}=\frac{1}{K}\frac{\partial u}{\partial t}, \tag{8.155}$$

and hence that, if T is finite (bounded) at the centre of the sphere and $\partial T/\partial r=0$ on the surface $r=a$ for $t>0$, the solution is of the form

$$T(r, t)=A+\frac{1}{r}\sum_{n=1}^{\infty} B_n e^{-K\omega_n^2 t}\sin(\omega_n r), \tag{8.156}$$

where A, B_n are constants, and ω_n $(n=1, 2, 3, \ldots)$ are determined by the positive roots of

$$\tan(\omega_n a)=\omega_n a. \tag{8.157}$$

To solve this problem, we use the form of ∇^2 in spherical polar coordinates (r, θ, ϕ) given in (1.84). We have, for $T=T(r, t)$,

$$\nabla^2 T=\frac{1}{r^2}\frac{\partial}{\partial r}\left(r^2\frac{\partial T}{\partial r}\right)=\frac{1}{K}\frac{\partial T}{\partial t}. \tag{8.158}$$

Hence

$$\frac{\partial^2 T}{\partial r^2} + \frac{2}{r}\frac{\partial T}{\partial r} = \frac{1}{K}\frac{\partial T}{\partial t}. \tag{8.159}$$

Now putting $u = rT$, we have

$$\frac{\partial T}{\partial r} = \frac{1}{r}\frac{\partial u}{\partial r} - \frac{1}{r^2}u \tag{8.160}$$

and

$$\frac{\partial^2 T}{\partial r^2} = \frac{1}{r}\frac{\partial^2 u}{\partial r^2} - \frac{2}{r^2}\frac{\partial u}{\partial r} + \frac{2}{r^3}u. \tag{8.161}$$

Substituting into (8.159), we find, as required by (8.155),

$$\frac{\partial^2 u}{\partial r^2} = \frac{1}{K}\frac{\partial u}{\partial t}, \tag{8.162}$$

and writing $u(r, t)$ in separable form as

$$u(r, t) = R(r)S(t), \tag{8.163}$$

we have

$$\frac{R''}{R} = \frac{1}{K}\frac{S'}{S} = \alpha, \tag{8.164}$$

where α is the separation constant. For convenience, we write $\alpha = \pm\omega^2$ to indicate that the separation constant may be of either sign. The solution of (8.164) for S with $\omega \neq 0$ is

$$S = A\mathrm{e}^{\pm K\omega^2 t}. \tag{8.165}$$

In order that the temperature falls off exponentially with time rather than increasing without limit (a non-physical solution), we must exclude the solution with the positive sign in (8.165). Hence the solutions of (8.164) for $\omega \neq 0$ are

$$R = B\cos(\omega r) + C\sin(\omega r), \quad S = A\mathrm{e}^{-K\omega^2 t}, \tag{8.166}$$

where A, B and C are constants. Hence, for $\omega \neq 0$,

$$u = [B\cos(\omega r) + C\sin(\omega r)]\mathrm{e}^{-K\omega^2 t}, \tag{8.167}$$

where the constant A has been absorbed into B and C. For $\omega = 0$,

$$S = D, \tag{8.168}$$

$$R = Er + F, \tag{8.169}$$

where D, E and F are constants, giving

$$u = Er + F, \tag{8.170}$$

where D has been absorbed into E and F.

We may now write the general solution as a superposition of the infinity of possible solutions, one for each value of ω ($\omega = 0$, and $\omega = \omega_1$, $\omega = \omega_2, \ldots$ which are yet to be found). Hence

$$u(r, t) = Er + F + \sum_{n=1}^{\infty} [B_n \cos(\omega_n r) + C_n \sin(\omega_r r)]e^{-K\omega_n^2 t}. \tag{8.171}$$

Accordingly, since $u = rT$, we have

$$T(r, t) = E + \frac{F}{r} + \frac{1}{r}\sum_{n=1}^{\infty} [B_n \cos(\omega_n r) + C_n \sin(\omega_n r)]e^{-K\omega_n^2 t}. \tag{8.172}$$

Now for T to be finite at the centre of the sphere, we must remove any term behaving as $1/r$ as $r \to 0$ since this will diverge. Hence $F = 0$, and $B_n = 0$ (for all n) since $\cos(\omega_n r)/r \to \infty$ as $r \to 0$. Consequently

$$T(r, t) = E + \frac{1}{r}\sum_{n=1}^{\infty} C_n \sin(\omega_n r)e^{-K\omega_n^2 t}, \tag{8.173}$$

and for $\partial T/\partial r = 0$ on $r = a$, we have

$$\left[\sum_{n=1}^{\infty} C_n e^{-K\omega_n^2 t} \frac{d}{dr}\left(\frac{\sin(\omega_n r)}{r}\right)\right]_{r=a} = 0, \tag{8.174}$$

or

$$[r\omega_n \cos(\omega_n r) - \sin(\omega_n r)]_{r=a} = 0, \tag{8.175}$$

giving

$$\tan(\omega_n a) = \omega_n a, \tag{8.176}$$

as required. An infinity of ω_n values satisfy this transcendental equation and these values must be used in (8.173) to form the solution $T(r, t)$.

By imposing some initial condition in the form $T(r, 0) = f(r)$, we may determine the constants E and C_n as follows. From (8.173), inserting $t = 0$, we have

$$T(r, 0) = f(r) = E + \frac{1}{r}\sum_{n=1}^{\infty} C_n \sin(\omega_n r). \tag{8.177}$$

Hence

$$rf(r) = Er + \sum_{n=1}^{\infty} C_n \sin(\omega_n r). \tag{8.178}$$

Multiplying by $\sin(\omega_m r)$ and integrating from $r = 0$ to $r = a$ gives

$$\int_0^a rf(r)\sin(\omega_m r)\,dr = E\int_0^a r\sin(\omega_m r)\,dr + \sum_{n=1}^{\infty}\int_0^a C_n \sin(\omega_n r)\sin(\omega_m r)\,dr. \quad (8.179)$$

Now integrating by parts and using (8.176), we find

$$\int_0^a r\sin(\omega_m r)\,dr = 0. \quad (8.180)$$

Further

$$\int_0^a \sin(\omega_n r)\sin(\omega_m r)\,dr = \frac{1}{2}\left(\frac{\omega_m^2 a^3}{1+\omega_m^2 a^2}\right)\delta_{mn}, \quad (8.181)$$

for $m, n = 1, 2, 3, \ldots$, using $\sin(\omega_n a) = \omega_n a/\surd(1+\omega_n^2 a^2)$ and $\cos(\omega_n a) = 1/\surd(1+\omega_n^2 a^2)$ from (8.176). Hence from (8.179) we find

$$\int_0^a rf(r)\sin(\omega_m r)\,dr = \frac{1}{2}\left(\frac{\omega_m^2 a^3}{1+\omega_n^2 a^2}\right)C_m, \quad (8.182)$$

which determines C_m in terms of an integral.

Similarly, multiplying (8.178) by r and integrating from $r = 0$ to $r = a$, we have

$$\int_0^a r^2 f(r)\,dr = E\int_0^a r^2\,dr + \sum_{n=1}^{\infty} C_n \int_0^a r\sin(\omega_n r)\,dr \quad (8.183)$$

$$= \frac{a^3 E}{3}, \quad (8.184)$$

using (8.180). Hence

$$E = \frac{3}{a^3}\int_0^a r^2 f(r)\,dr. \quad (8.185)$$

Equations (8.182) and (8.185) determine the constants in the solution (8.173) in terms of the given function $f(r)$. ◢

In many problems, the solution of a partial differential equation obtained using the separation of variables technique involves some of the special functions discussed in Chapter 2. The concluding example of this section is of this type, and concerns the Schrödinger equation of quantum mechanics (see (8.12)).

Example 10 To solve

$$\nabla^2 u + \alpha E u = 0, \tag{8.186}$$

where α is a constant and E is the energy parameter, subject to the conditions that u is a single-valued function and is bounded (finite) everywhere within a sphere of radius a, given $u = 0$ on $r = a$.

Expressing ∇^2 in spherical polar coordinates (r, θ, ϕ), using (1.84), (8.186) becomes

$$\frac{1}{r^2}\frac{\partial}{\partial r}\left(r^2\frac{\partial u}{\partial r}\right) + \frac{1}{r^2 \sin\theta}\frac{\partial}{\partial\theta}\left(\sin\theta\frac{\partial u}{\partial\theta}\right) + \frac{1}{r^2\sin^2\theta}\frac{\partial^2 u}{\partial\phi^2} + \alpha E u = 0. \tag{8.187}$$

Writing

$$u(r, \theta, \phi) = R(r)\Theta(\theta)\Phi(\phi), \tag{8.188}$$

we have

$$\frac{1}{Rr^2}\frac{d}{dr}\left(r^2\frac{dR}{dr}\right) + \frac{1}{r^2\sin\theta}\frac{1}{\Theta}\frac{d}{d\theta}\left(\sin\theta\frac{d\Theta}{d\theta}\right) + \frac{1}{r^2\sin^2\theta}\frac{1}{\Phi}\frac{d^2\Phi}{d\phi^2} + \alpha E = 0. \tag{8.189}$$

Letting

$$\frac{1}{\Phi}\frac{d^2\Phi}{d\phi^2} = -m^2, \tag{8.190}$$

where m is a constant, we have

$$\Phi = Ae^{im\phi}, \tag{8.191}$$

which, provided $m = 0, \pm 1, \pm 2, \ldots$ is such that for every rotation of 2π we return to the same value of Φ. This is required by the condition of single-valuedness of the solution for u. Using (8.190) in (8.189), we find

$$\frac{1}{R}\frac{d}{dr}\left(r^2\frac{dR}{dr}\right) + \alpha E r^2 = -\left[\frac{1}{\Theta\sin\theta}\frac{d}{d\theta}\left(\sin\theta\frac{d\Theta}{d\theta}\right) - \frac{m^2}{\sin^2\theta}\right]. \tag{8.192}$$

The terms in r and θ have now been separated so we may write

$$\frac{1}{\Theta\sin\theta}\frac{d}{d\theta}\left(\sin\theta\frac{d\Theta}{d\theta}\right) - \frac{m^2}{\sin^2\theta} = \lambda, \tag{8.193}$$

where λ is the separation constant. This equation is discussed in Chapter 2 (see (2.189)) where it is noted that bounded solutions exist only if

$$\lambda = -l(l+1), \quad l = 0, 1, 2, \ldots, \tag{8.194}$$

and that these solutions are given in terms of the associated Legendre functions

$$\Theta = BP_l^{|m|}(\cos\theta), \tag{8.195}$$

where $|m| \leqslant l$, and B is an arbitrary constant. Finally we solve the equation for $R(r)$ which, from (8.192) and (8.193), is

$$\frac{1}{r^2}\frac{d}{dr}\left(r^2\frac{dR}{dr}\right) + \left[\alpha E - \frac{l(l+1)}{r^2}\right]R = 0. \tag{8.196}$$

This equation may be put into a recognisable form by the substitutions

$$\sqrt{(\alpha E)}r = s, \quad R(r) = \frac{1}{\sqrt{s}}P(s), \tag{8.197}$$

giving

$$s^2\frac{d^2P}{ds^2} + s\frac{dP}{ds} + [s^2 - (l+\tfrac{1}{2})^2]P = 0. \tag{8.198}$$

This is Bessel's equation of order $\nu = l + \frac{1}{2}$ (see (2.164)) which has the bounded solution

$$P(s) = J_{l+\frac{1}{2}}(s), \tag{8.199}$$

the second solution $Y_{l+\frac{1}{2}}(s)$ being unbounded as $s \to 0$. Hence

$$R(r) = \frac{C}{\sqrt{r}}J_{l+\frac{1}{2}}(\sqrt{(\alpha E)}r). \tag{8.200}$$

Combining the results (8.191), (8.195) and (8.200), we finally have

$$u(r, \theta, \phi) = Ke^{im\phi}P_l^{|m|}(\cos\theta)\frac{1}{\sqrt{r}}J_{l+\frac{1}{2}}(\sqrt{(\alpha E)}r), \tag{8.201}$$

where K is an arbitrary constant which can be determined by imposing a further condition on u. In quantum mechanics this requirement is that

$$\iiint_{\text{all space}} u^*ur^2\sin\theta\,dr\,d\theta\,d\phi = 1, \tag{8.202}$$

thus normalising $u^*u = |u|^2$ to unity when integrated over all space. Now if $u = 0$ on $r = a$, we have

$$J_{l+\frac{1}{2}}(\sqrt{(\alpha E)}a) = 0. \tag{8.203}$$

For each given l value, there will be an infinity of roots $r_1, r_2, r_3, \ldots$ such that $J_{l+\frac{1}{2}}(r_i) = 0$ (the values of the r_i being the zeros of the Bessel function). Hence

$$\alpha E a^2 = r_1^2, r_2^2, r_3^2, \ldots, \tag{8.204}$$

and the energy E therefore takes on an infinite set of discrete values (this property being known as the quantisation of energy). ◢

We conclude this section by remarking that the examples given here are only representative of a whole variety of problems which can be solved in this way. For reasons of space, we do not discuss non-homogeneous equations or non-homogeneous boundary conditions (but see Problem 9 of this chapter).

8.8 Integral transform techniques

The integral transforms discussed in Chapter 7 are useful in solving a variety of partial differential equations, the choice of the most appropriate transform depending on the type of boundary or initial conditions to be imposed on the equation. The following examples illustrate the use of the Laplace and Fourier transforms.

1. Laplace transforms

Example 11 To solve the diffusion equation

$$\frac{\partial^2 u}{\partial x^2} = \frac{1}{K}\frac{\partial u}{\partial t}, \tag{8.205}$$

for $x > 0$, $t > 0$ given that $u = u_0$ (a constant) on $x = 0$ for $t > 0$, and $u = 0$ for $x > 0$, $t = 0$.

Taking the Laplace transform with respect to t gives

$$\frac{d^2\bar{u}(x, s)}{dx^2} = \frac{1}{K} s\bar{u}(x, s), \tag{8.206}$$

using (7.86) with $u(x, 0) = 0$. Also

$$\bar{u}(0, s) = \int_0^\infty u(0, t)\mathrm{e}^{-st}\,dt = u_0/s. \tag{8.207}$$

Hence solving (8.206) gives

$$\bar{u}(x, s) = A(s)\mathrm{e}^{-x\sqrt{(s/K)}} + B(s)\mathrm{e}^{x\sqrt{(s/K)}}. \tag{8.208}$$

To exclude solutions becoming infinite as $x \to \infty$, we take $B(s) = 0$. Then using (8.207) we find

$$A(s) = u_0/s \tag{8.209}$$

and hence

$$\bar{u}(x, s) = \frac{u_0}{s} e^{-x\sqrt{(s/K)}}. \tag{8.210}$$

The inversion of this function with respect to t is a standard result which we obtained in Section 7.5. From (7.57) with $a = x/\sqrt{K}$ we therefore have

$$u(x, t) = u_0 \mathscr{L}^{-1}\left\{\frac{1}{s} e^{-x\sqrt{(s/K)}}\right\} = u_0 \operatorname{erfc}\left[\frac{x}{2\sqrt{(Kt)}}\right], \tag{8.211}$$

where erfc is defined in (2.49). ◢

Example 12 To solve the equation

$$\frac{\partial^2 u}{\partial t^2} = \frac{\partial^2 u}{\partial x^2} + K^2 \frac{\partial^4 u}{\partial x^2 \partial t^2} \tag{8.212}$$

for $x > 0$, $t > 0$ (K being a real constant) given that

$$u = \partial u/\partial t = 0 \quad \text{for} \quad t = 0,\ x > 0, \tag{8.213}$$

$$u = 1 \quad \text{for} \quad x = 0,\ t > 0, \tag{8.214}$$

$$u \to 0 \quad \text{as} \quad x \to \infty,\ t > 0, \tag{8.215}$$

and to show that

$$\left(\frac{\partial u}{\partial x}\right)_{x=0} = -\frac{1}{K} J_0(t/K). \tag{8.216}$$

We recall that the Laplace transform with respect to t of time-derivatives of u involve precisely the values of u and $\partial u/\partial t$ at $t = 0$ which are given in (8.213). Hence taking the Laplace transform with respect to t is an appropriate method of solving the problem.

Using the results of Chapter 7 (see (7.87) and (7.90)), (8.212) becomes

$$s^2\bar{u}(x, s) - su(x, 0) - \left(\frac{\partial u}{\partial t}\right)_{t=0}$$
$$= \frac{d^2\bar{u}}{dx^2} + K^2 \frac{d^2}{dx^2}\left[s^2\bar{u}(x, s) - su(x, 0) - \left(\frac{\partial u}{\partial t}\right)_{t=0}\right]. \tag{8.217}$$

From (8.213), $u(x, 0) = (\partial u/\partial t)_{t=0} = 0$. Hence (8.217) becomes

$$\frac{d^2\bar{u}}{dx^2} = \frac{s^2}{1 + K^2 s^2} \bar{u}, \tag{8.218}$$

where $\bar{u} = \bar{u}(x, s)$. Solving this equation gives

$$\bar{u}(x, s) = A(s)e^{-sx/\sqrt{(1+K^2s^2)}} + B(s)e^{sx/\sqrt{(1+K^2s^2)}}. \tag{8.219}$$

Since $u(x, t) \to 0$ as $x \to \infty$, it follows that $\bar{u}(x, s) \to 0$ as $x \to \infty$ and we must therefore have $B(s) = 0$. Accordingly

$$\bar{u}(x, s) = A(s)e^{-sx/\sqrt{(1+K^2s^2)}}. \tag{8.220}$$

Now $u(0, t) = 1$ by (8.214), so that

$$\bar{u}(0, s) = \int_0^\infty 1 . e^{-st}\, dt = 1/s. \tag{8.221}$$

Therefore in (8.220), $A(s) = 1/s$ and

$$\bar{u}(x, s) = \frac{1}{s} e^{-sx/\sqrt{(1+K^2s^2)}}. \tag{8.222}$$

Using the Bromwich contour (see Section 7.8) to evaluate this integral, we have the solution in the form

$$u(x, t) = \frac{1}{2\pi i} \int_{\gamma - i\infty}^{\gamma + i\infty} \frac{1}{s} e^{-sx/\sqrt{(1+K^2s^2)}} e^{st}\, ds. \tag{8.223}$$

From (8.223)

$$\left(\frac{\partial u}{\partial x}\right)_{x=0} = -\frac{1}{2\pi i}\left[\int_{\gamma - i\infty}^{\gamma + i\infty} \frac{e^{-sx/\sqrt{(1+K^2s^2)}}}{\sqrt{(1+K^2s^2)}} e^{st}\, ds\right]_{x=0} \tag{8.224}$$

$$= -\frac{1}{2\pi i} \int_{\gamma - i\infty}^{\gamma + i\infty} \frac{e^{st}}{\sqrt{(1+K^2s^2)}}\, ds \tag{8.225}$$

$$= -\frac{1}{2\pi i K} \int_{\gamma' - i\infty}^{\gamma' + i\infty} \frac{e^{s't/K}}{\sqrt{(1+s'^2)}}\, ds', \tag{8.226}$$

where $s' = Ks$. Since the Laplace inversion of $1/\sqrt{(1+s^2)}$ is $J_0(x)$ (see Chapter 7, Example 11), (8.226) gives

$$\left(\frac{\partial u}{\partial x}\right)_{x=0} = -\frac{1}{K} J_0(t/K). \tag{8.227}$$ ◢

2. *Fourier transforms*

Finally we illustrate the use of the Fourier transform in solving particular types of boundary value problems. In Chapter 7, it was found that the Fourier sine transform of a second derivative term gave the result

$$\int_0^\infty \frac{\partial^2 u}{\partial x^2} \sin(sx)\, dx = -s^2 \bar{u}_S(s, t) + su(0, t), \tag{8.228}$$

whereas the cosine transform gave

$$\int_0^\infty \frac{\partial^2 u}{\partial x^2}\cos(sx)\,dx = -s^2\bar{u}_C(s,t) - \left(\frac{\partial u}{\partial x}\right)_{x=0}, \tag{8.229}$$

$\bar{u}_S$ and $\bar{u}_C$ being respectively the Fourier sine and cosine transforms of $u(x, t)$ with respect to x. The choice of a suitable transform is therefore dependent on the nature of the given boundary conditions. If $u(0, t)$ is specified, but not $(\partial u/\partial x)_{x=0}$, then the sine transform is appropriate, whereas if $(\partial u/\partial x)_{x=0}$ is specified but not $u(0, t)$ then the cosine transform is appropriate. We now give an example.

Example 13 To show, using the basic result

$$\int_0^\infty e^{-u^2}\cos(\lambda u)\,du = \frac{\sqrt{\pi}}{2}e^{-\lambda^2/4}, \tag{8.230}$$

that the Fourier sine transform of $\operatorname{erfc}(ax)$ with respect to x is $(1 - e^{-s^2/4a^2})/s$, and hence to solve

$$\frac{\partial^2 u}{\partial x^2} = \frac{1}{K}\frac{\partial u}{\partial t} \tag{8.231}$$

for $x > 0$, $t > 0$ given that $u = u_0$ (a constant) on $x = 0$ for $t > 0$, and $u = 0$ for $x > 0$, $t = 0$. (We note that this is the same problem as in Example 11 which we now solve by taking the Fourier transform with respect to x instead of taking the Laplace transform with respect to t.)

Using the definition (2.49),

$$\operatorname{erfc}(ax) = \frac{2}{\sqrt{\pi}}\int_{ax}^\infty e^{-u^2}\,du. \tag{8.232}$$

Hence its Fourier sine transform is

$$\mathscr{F}_S\{\operatorname{erfc}(ax)\} = \frac{2}{\sqrt{\pi}}\int_0^\infty \sin(sx)\int_{ax}^\infty e^{-u^2}\,du\,dx. \tag{8.233}$$

Inverting the order of integration, we have

$$\mathscr{F}_S\{\operatorname{erfc}(ax)\} = \frac{2}{\sqrt{\pi}}\int_0^\infty e^{-u^2}\,du\int_0^{u/a}\sin(sx)\,dx \tag{8.234}$$

$$= \frac{2}{\sqrt{\pi}}\int_0^\infty \frac{e^{-u^2}}{s}\left[1 - \cos\left(\frac{us}{a}\right)\right]du \tag{8.235}$$

$$= \frac{2}{s\sqrt{\pi}}\left(\frac{\sqrt{\pi}}{2} - \frac{\sqrt{\pi}}{2}e^{-s^2/4a^2}\right) \tag{8.236}$$

$$= \frac{1}{s}(1 - e^{-s^2/4a^2}), \tag{8.237}$$

using (8.230).

Since $u(0, t) = u_0$ is specified, the Fourier sine transform is appropriate in this case. From (8.231), using (8.228), we therefore have

$$-s^2\bar{u}_S(s, t) + su(0, t) = \frac{1}{K}\frac{d\bar{u}_S(s, t)}{dt}, \tag{8.238}$$

or

$$\frac{d\bar{u}_S}{dt} + Ks^2\bar{u}_S = Ksu_0. \tag{8.239}$$

The solution of this equation is easily found to be

$$\bar{u}_S(s, t) = A(s)e^{-Ks^2t} + u_0/s. \tag{8.240}$$

Since $u(x, 0) = 0$, we have $\bar{u}(s, 0) = 0$. From (8.240) it follows that

$$0 = A(s) + u_0/s, \tag{8.241}$$

which determines $A(s)$. Hence

$$\bar{u}_S(s, t) = (u_0/s)(1 - e^{-Ks^2t}). \tag{8.242}$$

Using the result (8.237), we obtain

$$u(x, t) = u_0 \operatorname{erfc}[x/2\sqrt{(Kt)}], \tag{8.243}$$

which was the solution found in Example 11. ◢

Problems 8

1. Use characteristics to solve

   (i) $y\dfrac{\partial u}{\partial x} + x\dfrac{\partial u}{\partial y} = 0$

   given $u = \cos x$ on $x^2 + y^2 = 1$,

   (ii) $x^2\dfrac{\partial u}{\partial x} + y^2\dfrac{\partial u}{\partial y} = 0$

   given $u \to e^y$ as $x \to \infty$,

   (iii) $x\dfrac{\partial u}{\partial x} - y\dfrac{\partial u}{\partial y} = u^2$

   given $u = 1/\ln x$ on $x^2y = 1$,

   (iv) $y\dfrac{\partial u}{\partial x} - 2xy\dfrac{\partial u}{\partial y} = 2xu$

   given $u(0, y) = \sinh y/y$.

2. Show that the equation

$$xy\frac{\partial^2\phi}{\partial x^2}+(y^2-x^2)\frac{\partial^2\phi}{\partial x\,\partial y}-xy\frac{\partial^2\phi}{\partial y^2}-x\frac{\partial\phi}{\partial y}+y\frac{\partial\phi}{\partial x}=0$$

can be written as $L_1L_2\phi=0$, where

$$L_1=x\frac{\partial}{\partial x}+y\frac{\partial}{\partial y},\quad L_2=y\frac{\partial}{\partial x}-x\frac{\partial}{\partial y}.$$

By writing $\psi=L_2\phi$, obtain by the method of characteristics the solution of the original equation subject to the conditions

$$\phi=\tfrac{1}{4}y^2 \quad\text{on}\quad x=1,\quad \partial\phi/\partial x=\tfrac{3}{2}\quad\text{on}\quad x=1.$$

3. Show that the equation

$$\frac{\partial^2 u}{\partial x^2}+4\frac{\partial u}{\partial x}+\frac{\partial^2 u}{\partial y^2}+4u=0$$

can be transformed into

$$\frac{\partial^2 v}{\partial x^2}+\frac{\partial^2 v}{\partial y^2}=0$$

by putting $u=\mathrm{e}^{-2x}v$. Hence obtain the general solution of the equation for u.

4. Solve

$$x\frac{\partial u}{\partial x}-2\frac{\partial u}{\partial y}=2x^2+2-4y$$

subject to the condition $u(1,y)=1+y^2+\mathrm{e}^y$.

5. By writing $\phi=\psi+\tfrac{1}{3}(x^2+y^2)$, solve the equation

$$x\frac{\partial\phi}{\partial x}+y\frac{\partial\phi}{\partial y}+\phi=x^2+y^2$$

subject to the boundary condition

$$\phi=\mathrm{e}^{-y^2}\cos y+\tfrac{1}{3}(1+y^2)\quad\text{on}\quad x=1.$$

6. Show that $v(x,t)=f(t)\mathrm{e}^{x^2g(t)}$ satisfies the diffusion equation

$$\frac{\partial^2 v}{\partial x^2}=\frac{\partial v}{\partial t}$$

if f and g are solutions of the equations

$$\frac{df}{dt}-2fg=0,\quad \frac{dg}{dt}-4g^2=0.$$

Hence show that if $v(0, t) = 1/\sqrt{(1+t)}$, then

$$v(x, t) = \frac{1}{\sqrt{(1+t)}} e^{-x^2/4(1+t)}.$$

7. Prove that if K is a positive constant, any solution $u = u(r, t)$ of the equation

$$\frac{\partial^2 u}{\partial r^2} + \frac{1}{r}\frac{\partial u}{\partial r} = \frac{1}{K}\frac{\partial u}{\partial t}$$

which is of the separable form $u(r, t) = R(r)T(t)$ and which tends to zero as $t \to \infty$ is of the form

$$u = e^{-K\lambda^2 t}[AJ_0(\lambda r) + BY_0(\lambda r)],$$

where A, B and λ are constants, and $J_0(x)$ and $Y_0(x)$ are the zero-order Bessel functions satisfying

$$xy'' + y' + xy = 0.$$

Given $u(a, t) = 0$ and $u(r, 0) = J_0(\omega r/a)$, where ω is defined by $J_0(\omega) = 0$, find the value of t such that

$$u(0, t) = \tfrac{1}{2}u(0, 0).$$

8. An infinite cylinder with circular cross-section has radius $a = 2$ units. The temperature distribution within the cylinder is $T(r, t)$. The surface of the cylinder is kept at zero temperature for all time, and at $t = 0$,

$$T(r, 0) = J_0(1.2r),$$

where J_0 is the Bessel function of zero order. Writing the heat conduction equation

$$\nabla^2 T = \frac{1}{K}\frac{\partial T}{\partial t}$$

in cylindrical coordinates, show that the temperature distribution is approximately

$$T(r, t) = e^{-1.44Kt}J_0(1.2r).$$

9. If the diffusion equation

$$\frac{\partial^2 u}{\partial x^2} = \frac{1}{K}\frac{\partial u}{\partial t}$$

is to be solved for $u(x, t)$ subject to the inhomogeneous boundary conditions

$$u(0, t) = U_0, \qquad t>0,$$
$$u(l, t) = U_1, \qquad t>0,$$

where U_0, U_1 and l are given constants, and

$$u(x, 0) = f(x), \quad 0 \leqslant x \leqslant l,$$

where $f(x)$ is a given function, show that the substitution

$$u(x, t) = v(x) + w(x, t)$$

can lead to a homogeneous boundary value problem for $w(x, t)$ by suitably choosing $v(x)$.

Show that the solution of the diffusion equation when $U_0 = 0$, $U_1 = 1$, $l = 1$ and $f(x) = 0$ is

$$u(x, t) = x + \frac{2}{\pi} \sum_{n=1}^{\infty} \frac{(-1)^n}{n} \sin(n\pi x) \mathrm{e}^{-n^2\pi^2 Kt}.$$

10. The function $u(r, t)$ satisfies the wave equation

$$\frac{1}{c^2}\frac{\partial^2 u}{\partial t^2} = \frac{\partial^2 u}{\partial r^2} + \frac{2}{r}\frac{\partial u}{\partial r}$$

within the sphere $0 \leqslant r < a$ where c is a constant and $t > 0$. Given

$$u = 0 \quad \text{and} \quad \partial u/\partial t = 1 \quad \text{at} \quad t = 0;$$
$$u = 0 \quad \text{on} \quad r = a;$$
$$u \text{ is finite as } r \to 0,$$

show, by taking the Laplace transform with respect to t, that

$$\bar{u}(r, s) = \mathscr{L}\{u(r, t)\} = \frac{1}{s^2}\left(1 - \frac{a \sinh(sr/c)}{r \sinh(sa/c)}\right).$$

Using the result

$$\mathscr{L}^{-1}\left\{\frac{1}{s^2}\frac{\sinh(sx)}{\sinh(sa)}\right\} = \frac{xt}{a} + \frac{2a}{\pi^2}\sum_{n=1}^{\infty}\frac{(-1)^n}{n}\sin\left(\frac{n\pi x}{a}\right)\sin\left(\frac{n\pi t}{a}\right),$$

obtain a series expansion for $u(r, t)$.

11. Show that, if $u(x, t)$ satisfies the differential equation

$$a\frac{\partial^2 u}{\partial x^2} + b\mathrm{e}^{-cx} = \frac{\partial u}{\partial t}, \quad (x \geqslant 0, t \geqslant 0),$$

where a, b and c are constants, and the boundary conditions are

$$u = 0 \quad \text{at} \quad t = 0 \quad \text{for all } x;$$
$$u \to 0 \quad \text{as} \quad x \to \infty \quad \text{for all } t;$$
$$\partial u/\partial x = 0 \quad \text{at} \quad x = 0 \quad \text{for all } t,$$

then

$$\mathscr{L}\{u(x, t)\} = \bar{u}(x, s) = \frac{b}{as(c^2 - s/a)}\left[\frac{c}{\sqrt{(s/a)}}\mathrm{e}^{-x\sqrt{(s/a)}} - \mathrm{e}^{-cx}\right].$$

Using the result

$$\mathscr{L}^{-1}\left\{\frac{1}{\sqrt{s}}\mathrm{e}^{-x\sqrt{(s/a)}}\right\} = \frac{1}{\sqrt{(\pi t)}}\mathrm{e}^{-x^2/4at},$$

find $u(x, t)$.

12. Show that the solution of the heat conduction equation

$$\frac{\partial^2 u}{\partial x^2} = \frac{1}{K}\frac{\partial u}{\partial t}, \quad (0 \leqslant x \leqslant l, t > 0),$$

subject to the conditions

$$\frac{\partial u}{\partial x} = 0 \quad \text{for } x = 0, \text{ all } t;$$
$$u(l, t) = u_0 \quad \text{for all } t,$$

has the Laplace transform

$$\bar{u}(x, s) = \frac{u_0}{s}\frac{\cosh[\sqrt{(s/K)}x]}{\cosh[\sqrt{(s/K)}l]}.$$

By finding the residues of $\bar{u}(x, s)\mathrm{e}^{st}$ at the poles $s = 0$ and $s = -(2n-1)^2(\pi^2 K/4l^2)$, where $n = 1, 2, 3, \ldots$, show that

$$u(x, t) = u_0\left\{1 + \frac{4}{\pi}\sum_{n=1}^{\infty}\frac{(-1)^n}{(2n-1)}\cos\left[\frac{(2n-1)\pi x}{2l}\right]\mathrm{e}^{-(2n-1)^2\pi^2 Kt/4l^2}\right\}.$$

13. Solve, using the Fourier transform, Helmholtz's equation

$$\nabla^2 u = u,$$

where $-\infty < x < \infty$, $0 \leqslant y \leqslant 1$, given

$$(\partial u/\partial y)_{y=0} = 0 \quad \text{for all } x;$$
$$u(x, 1) = \mathrm{e}^{-\alpha x^2} \quad (\alpha \text{ positive}).$$

14. Show that the solutions to Burger's equation

$$\frac{\partial u}{\partial t}+u\frac{\partial u}{\partial x}=\nu\frac{\partial^2 u}{\partial x^2} \qquad (\nu=\text{constant})$$

may be obtained from solutions of $\partial\theta/\partial t=\nu\,\partial^2\theta/\partial x^2$ by making u proportional to $(\partial/\partial x)(\ln\theta)$. Derive this constant of proportionality.

Use this transformation, followed by separation of variables for the linear θ-equation, to solve Burger's equation for $0\leqslant x\leqslant a$, $t>0$ with the initial condition

$$u(x,0)=u_0, \quad (0\leqslant x\leqslant a),$$

where u_0 is a constant, and the boundary conditions

$$u(0,t)=u(a,t)=0, \quad (t>0).$$

9
Calculus of variations

9.1 Introduction

One of the elementary uses of differential calculus is in finding the stationary values of functions of one or more variables. The calculus of variations considers the more complicated problem of finding the stationary values of integrals. A simple physical example which illustrates the general type of mathematical problem involved is as follows: suppose $P(x_1, y_1)$ and $Q(x_2, y_2)$ are two given fixed points in a cartesian coordinate system (see Figure 9.1). We wish to determine the equation of the curve joining these two points such that when the curve is rotated through 2π about the x-axis to form a surface, the surface area so generated has a minimum value. Suppose $y = y(x)$ is some curve joining P and Q. Then, if ds is the element of arc length, the surface area S so formed is given by

$$S = 2\pi \int_P^Q y(x)\, ds = 2\pi \int_{x_1}^{x_2} y\sqrt{[1 + (dy/dx)^2]}\, dx. \tag{9.1}$$

S is not a function in the usual sense since it depends on the form of the curve $y = y(x)$ in the whole range $x_1 \leqslant x \leqslant x_2$. It is commonly called a functional and is written as $S[y(x)]$, the square brackets signifying that if the form of $y(x)$ is known then the value of S is known.

It is relatively straightforward to find the conditions under which (9.1) takes on a stationary value, but not usually easy to determine whether this is a maximum or a minimum, or some other type of stationary value. In many problems, however, the solution may be seen by physical arguments to be either a maximum or a minimum, and accordingly we shall not attempt here to give the detailed mathematical analysis determining the nature of the stationary values.

9.2 Euler's equation

The simplest problem is to find a function $y(x)$ such that the integral

$$I[y(x)] = \int_{x_1}^{x_2} f(x, y, y')\, dx \tag{9.2}$$

is stationary with respect to small changes in $y(x)$. Here y' denotes dy/dx, and x_1 and x_2 are given fixed limits. We see that, in the case of (9.1), the function $f(x, y, y')$ has the form

$$f(x, y, y') = 2\pi y \sqrt{(1 + y'^2)}. \tag{9.3}$$

Suppose $y = y(x)$ is some curve joining the points $P(x_1, y_1)$ and $Q(x_2, y_2)$ as in Figure 9.1. Consider a small variation in $y(x)$ so that the new curve has the form

$$y(x, \epsilon) = y(x) + \epsilon\eta(x), \tag{9.4}$$

where ϵ is a small parameter and $\eta(x)$ is an arbitrary function chosen so that

$$\eta(x_1) = \eta(x_2) = 0. \tag{9.5}$$

In this way the new curve (9.4) also passes through the end points P and Q. The term $\epsilon\eta(x)$ is denoted by δy and is called the variation of y. This variation may be one of two types – weak or strong – in the following sense. Suppose that the curve C_0 represents $y = y(x)$ (Figure 9.2). The neighbouring curve C_1 corresponds to a weak variation since

Figure 9.1

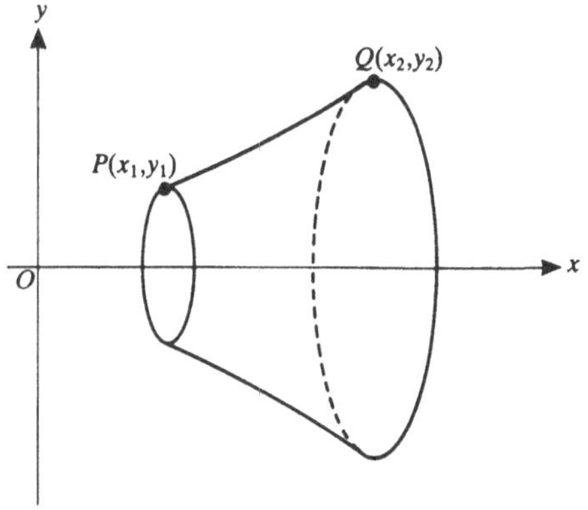

the difference between the gradients of C_0 and C_1, at any value of x in the range $x_1 \leqslant x \leqslant x_2$, is a small quantity. In other words

$$\delta y' = \frac{d}{dx}(\delta y) \tag{9.6}$$

is small. However, the curve C_2 (continuous but with a discontinuous derivative) corresponds to a strong variation since, although δy is itself small, the derivative $\delta y'$ is no longer small due to the abrupt changes in the slope of the curve. In most cases we deal only with weak variations, and assume that the curve $y(x) + \delta y$ is similar to C_1 and is differentiable to all desired orders.

Inserting (9.4) into (9.2) we can write

$$\begin{aligned} I(\epsilon) &= I[y(x) + \epsilon\eta(x)] \\ &= \int_{x_1}^{x_2} f(x, y(x) + \epsilon\eta(x), y'(x) + \epsilon\eta'(x))\,dx, \end{aligned} \tag{9.7}$$

whence

$$\begin{aligned} \delta I(\epsilon) &= I(\epsilon) - I \\ &= \int_{x_1}^{x_2} [f(x, y + \epsilon\eta, y' + \epsilon\eta') - f(x, y, y')]\,dx. \end{aligned} \tag{9.8}$$

Assuming now that the first term of the integrand may be expanded in powers of ϵ, we have (using Taylor's expansion of a function of two variables)

$$\delta I(\epsilon) = \epsilon I_1 + \frac{\epsilon^2}{2!} I_2 + \text{terms in } \epsilon^3 \text{ and higher orders,} \tag{9.9}$$

where

$$I_1 = \int_{x_1}^{x_2} \left(\eta \frac{\partial f}{\partial y} + \eta' \frac{\partial f}{\partial y'}\right) dx \tag{9.10}$$

Figure 9.2

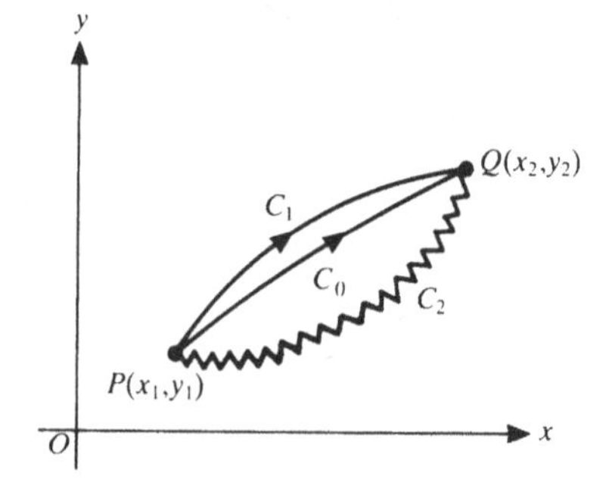

and

$$I_2 = \int_{x_1}^{x_2} \left(\eta^2 \frac{\partial^2 f}{\partial y^2} + 2\eta\eta' \frac{\partial^2 f}{\partial y\, \partial y'} + \eta'^2 \frac{\partial^2 f}{\partial y'^2} \right) dx. \tag{9.11}$$

To the first order in ϵ, the variation $\delta I(\epsilon)$ is given by

$$\delta I(\epsilon) = \epsilon I_1, \tag{9.12}$$

and the sign of $\delta I(\epsilon)$ therefore depends on the sign of ϵ. The value of $I[y(x)]$ in (9.2) therefore may increase for some variations and decrease for others. However, for the integral to have a maximum or minimum, all variations with respect to ϵ should respectively decrease or increase its value. This can be achieved by requiring in (9.12) that $I_1 = 0$, that is

$$\int_{x_1}^{x_2} \left(\eta \frac{\partial f}{\partial y} + \eta' \frac{\partial f}{\partial y'} \right) dx = 0. \tag{9.13}$$

Integrating the second term in (9.13) by parts we have

$$\int_{x_1}^{x_2} \eta' \frac{\partial f}{\partial y'}\, dx = \left[\eta \frac{\partial f}{\partial y'} \right]_{x_1}^{x_2} - \int_{x_1}^{x_2} \eta \frac{d}{dx}\left(\frac{\partial f}{\partial y'} \right) dx. \tag{9.14}$$

However $\eta(x_1) = \eta(x_2) = 0$ (see (9.5)) and the first term on the right of (9.14) is therefore zero. Inserting (9.14) into (9.13) gives

$$\int_{x_1}^{x_2} \eta(x) \left[\frac{\partial f}{\partial y} - \frac{d}{dx}\left(\frac{\partial f}{\partial y'} \right) \right] dx = 0. \tag{9.15}$$

Now $\eta(x)$ is an arbitrary function, apart from the conditions (9.5). Hence we may make use of a fundamental result which states that if $\eta(x)$ is continuous and has continuous derivatives up to at least second order in the range $x_1 \leq x \leq x_2$ with $\eta(x_1) = \eta(x_2) = 0$ and $\eta(x) \neq 0$ for $x_1 < x < x_2$, and if $g(x)$ is itself a continuous function and

$$\int_{x_1}^{x_2} \eta(x) g(x)\, dx = 0 \tag{9.16}$$

for arbitrary (every) $\eta(x)$ satisfying the conditions above, then $g(x) = 0$. Since (9.15) holds for arbitrary $\eta(x)$, it follows from the above fundamental result that a necessary condition for the integral to be stationary is

$$\frac{\partial f}{\partial y} - \frac{d}{dx}\left(\frac{\partial f}{\partial y'} \right) = 0. \tag{9.17}$$

This equation is known as Euler's equation and its solution leads to the form of $y(x)$ which produces the stationary value of I. The curves $y = y(x)$ so obtained are called extremals of (9.2).

Example 1 To find the stationary value of

$$I[y(x)] = \int_P^Q \left[\left(\frac{dy}{dx} \right)^2 - 2yx + y^2 \right] dx, \tag{9.18}$$

where $P = (0, 0)$ and $Q = (1, 1)$.

From (9.18)

$$f(x, y, y') = y'^2 - 2yx + y^2 \tag{9.19}$$

and hence

$$\frac{\partial f}{\partial y} = -2x + 2y, \quad \frac{\partial f}{\partial y'} = 2y'. \tag{9.20}$$

The Euler equation (9.17) is therefore

$$-2x + 2y - \frac{d}{dx}(2y') = 0 \tag{9.21}$$

or

$$\frac{d^2y}{dx^2} - y = -x. \tag{9.22}$$

The solution of this equation is

$$y = A \cosh x + B \sinh x + x. \tag{9.23}$$

By requiring this curve to pass through $P(0, 0)$, we have

$$A = 0, \tag{9.24}$$

while for it to pass through $Q(1, 1)$ we must have

$$B = 0. \tag{9.25}$$

Hence

$$y = x \tag{9.26}$$

is the extremal curve.

To evaluate the stationary value of (9.18), we insert (9.26) into the integrand of (9.18) and integrate. Then

$$I = \int_0^1 (1 - 2x^2 + x^2)\, dx \tag{9.27}$$

$$= \int_0^1 (1 - x^2)\, dx = \tfrac{2}{3}. \tag{9.28}$$ ◢

9.3 Alternative forms of Euler's equation

Euler's equation (9.17) assumes simpler forms when one or more of the variables x, y or y' are absent from the function f.

Case 1 If y is explicitly absent from f then

$$f(x, y, y') = f(x, y') \tag{9.29}$$

and hence $\partial f/\partial y = 0$. The Euler equation (9.17) now becomes

$$\frac{d}{dx}\left(\frac{\partial f}{\partial y'}\right) = 0, \tag{9.30}$$

giving

$$\partial f/\partial y' = \text{constant}. \tag{9.31}$$

Example 2 Suppose we want to find the equation of the curve of shortest length which joins two points P and Q in the plane. If $y = y(x)$ is some curve joining P to Q, then the element of arc length along the curve is given by

$$ds = \sqrt{(1 + y'^2)}\, dx. \tag{9.32}$$

Hence the total length of the curve from P to Q is

$$S = \int_P^Q \sqrt{(1 + y'^2)}\, dx. \tag{9.33}$$

We can now apply the Euler equation to this integral and, since the integrand is explicitly independent of y, we can use the form (9.31). The Euler equation therefore has the form

$$\frac{\partial}{\partial y'}[\sqrt{(1 + y'^2)}] = \text{constant} \tag{9.34}$$

or

$$\frac{y'}{\sqrt{(1 + y'^2)}} = C. \tag{9.35}$$

Hence

$$y' = A, \tag{9.36}$$

where A is a constant, and the equation of the curve of shortest length joining P and Q is therefore

$$y = Ax + B, \tag{9.37}$$

the constants A and B chosen so that (9.37) passes through the end points P and Q. ◢

Case 2 If y' is explicitly absent from f, then $f(x, y, y') = f(x, y)$. Therefore $\partial f/\partial y' = 0$ and the Euler equation (9.17) becomes

$$\partial f/\partial y = 0. \tag{9.38}$$

Example 3 To find the extremal curve of the integral

$$I = \int_0^1 (2y^4 - xy)\, dx, \tag{9.39}$$

we use (9.38) to obtain

$$\partial f/\partial y = 8y^3 - x = 0 \tag{9.40}$$

or

$$y = \tfrac{1}{2}x^{\frac{1}{3}}. \tag{9.41}$$

The stationary value of I, found by inserting (9.41) into (9.39) and integrating between the given limits, is $I = -\frac{9}{56}$. ◢

Case 3 If x is explicitly absent from f then $f(x, y, y') = f(y, y')$ and hence $\partial f/\partial x = 0$. To see the effect this has on the Euler equation (9.17), we note that the total differential of $f(x, y, y')$ with respect to x is given by

$$\frac{df}{dx} = \frac{\partial f}{\partial x} + \frac{\partial f}{\partial y}\frac{dy}{dx} + \frac{\partial f}{\partial y'}\frac{d}{dx}(y') \tag{9.42}$$

$$= y'\frac{\partial f}{\partial y} + y''\frac{\partial f}{\partial y'}, \tag{9.43}$$

using $\partial f/\partial x = 0$.

Now multiplying Euler's equation (9.17) by y', we have

$$y'\frac{\partial f}{\partial y} - y'\frac{d}{dx}\left(\frac{\partial f}{\partial y'}\right) = 0. \tag{9.44}$$

Eliminating $y'\, \partial f/\partial y$ between (9.43) and (9.44) gives

$$\frac{df}{dx} - y''\frac{\partial f}{\partial y'} - y'\frac{d}{dx}\left(\frac{\partial f}{\partial y'}\right) = 0 \tag{9.45}$$

or

$$\frac{d}{dx}\left(f - y'\frac{\partial f}{\partial y'}\right) = 0. \tag{9.46}$$

Integrating (9.46) we find

$$f - y'\frac{\partial f}{\partial y'} = C, \tag{9.47}$$

where C is a constant.

The following two examples illustrate the use of this form of the Euler equation.

Example 4 An ancient problem is to find the curve joining two given points in a vertical plane for which the time of descent of a particle down the curve (under a constant gravitational field) is a minimum. Such a curve is called the 'curve of quickest descent' or the brachistochrone (from two Greek words – *brachistos* meaning 'shortest' and *chronos* meaning 'time'). Suppose a particle starts from rest at a point O (see Figure 9.3) and slides down the curve OA, frictional forces being neglected. Then at a typical point $P(x, y)$, the velocity of the particle is

$$v = \sqrt{(2gy)}, \tag{9.48}$$

where g is the acceleration due to gravity. The time δt taken to travel from P to a neighbouring point P' is therefore

$$\delta t = \delta s/\sqrt{(2gy)}, \tag{9.49}$$

where δs is the length P to P'. Allowing P' to tend to P, we have

$$dt = \frac{ds}{\sqrt{(2gy)}} = \frac{\sqrt{(1+y'^2)}}{\sqrt{(2gy)}}\,dx, \tag{9.50}$$

and therefore the total time T from $O(0, 0)$ to $A(x_1, y_1)$ is

$$T = \frac{1}{\sqrt{(2g)}}\int_0^{x_1}\frac{\sqrt{(1+y'^2)}}{\sqrt{y}}\,dx. \tag{9.51}$$

We now wish to find $y(x)$ such that T is a minimum. The integrand of (9.51) is explicitly independent of x and hence we may use the form

Figure 9.3

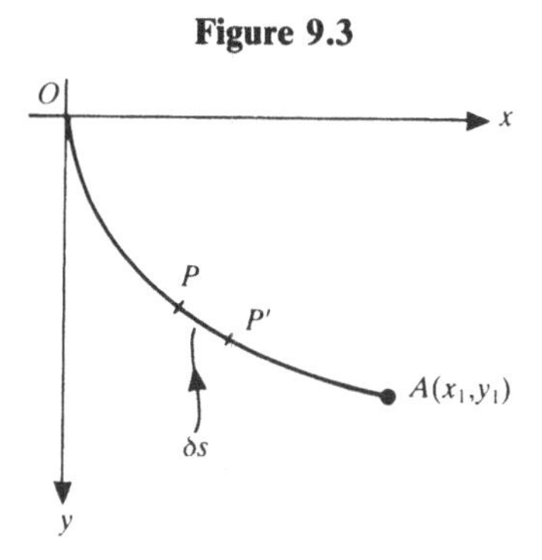

(9.47) of the Euler equation. Accordingly, we have

$$\frac{\sqrt{(1+y'^2)}}{\sqrt{y}} - y'\frac{\partial}{\partial y'}\left[\frac{\sqrt{(1+y'^2)}}{\sqrt{y}}\right] = C, \tag{9.52}$$

or

$$\sqrt{\left(\frac{1+y'^2}{y}\right)} - \frac{y'^2}{\sqrt{y}\sqrt{(1+y'^2)}} = C. \tag{9.53}$$

This simplifies to

$$y(1+y'^2) = D, \tag{9.54}$$

where D is a constant.

Making the substitution $y' = \cot\theta$, we have

$$y = \frac{D}{1+y'^2} = D\sin^2\theta = \frac{D}{2}[1-\cos(2\theta)] \tag{9.55}$$

and

$$\frac{dx}{d\theta} = \frac{1}{y'}\frac{dy}{d\theta} = D\tan\theta\sin(2\theta) = D[1-\cos(2\theta)], \tag{9.56}$$

whence

$$x = \frac{D}{2}[2\theta - \sin(2\theta)]. \tag{9.57}$$

Equations (9.55) and (9.57) give the parametric form of the equation of the curve, and define a cycloid. ◢

Example 5 We now return to the problem discussed in Section 9.1 of the generation of a surface of minimum area. The integral to be minimised is given in (9.1) and has an integrand which is explicitly independent of x. Hence, using the form (9.47) of Euler's equation, we obtain

$$y\sqrt{(1+y'^2)} - \frac{yy'^2}{\sqrt{(1+y'^2)}} = C. \tag{9.58}$$

After simplification, this becomes

$$y^2 = C^2(1+y'^2), \tag{9.59}$$

which integrates to give

$$y = C\cosh\left(\frac{x+A}{C}\right), \tag{9.60}$$

where the constants A and C must be chosen so that the curve passes through the end points P and Q. This curve is known as a catenary, and the surface of minimum area generated by this curve is known as a catenoid. ◢

9.4 Extremal curves with discontinuous derivatives

When the extremal is found to consist of two or more branches, we must check whether it is possible for the given end points to lie on one and the same part of the curve. If this is not possible then the curve is not the required extremal since the end points are not joined by a continuous curve. We can illustrate this by the following example.

Example 6 Suppose we require the curve $y = y(x)$ which minimises

$$I = \int_P^Q x^2 y'^2 \, dx, \tag{9.61}$$

where $P = (1, 2)$ and $Q = (-2, -1)$.

The Euler equation, from (9.31), is

$$2x^2 y' = C, \tag{9.62}$$

from which, by integration, we have

$$y = A/x + B. \tag{9.63}$$

The curve on which P and Q lie is therefore $y = 2/x$, but P is one branch of this curve and Q is on the other branch (see Figure 9.4). Hence this is not the required extremal (in fact, evaluating I with this curve gives infinity), and no extremal curve of the form (9.63) exists. However, provided extremal curves with discontinuous derivatives are allowed, we may obtain an extremal curve joining P to Q by noting that the integrand in (9.61) is a perfect square and hence its minimum value is zero. This is achieved by taking $y' = 0$, or $y = \text{constant}$. A possible extremal is shown in Figure 9.4, for which $y' = 0$ except at $x = 0$. ◢

9.5 Several dependent variables

The integral (9.2) and its associated Euler equation (9.17) may be generalised to n dependent variables $y_1(x), y_2(x), \ldots, y_n(x)$ by writing

$$I[y_1(x), y_2(x), \ldots, y_n(x)]$$
$$= \int_{x_1}^{x_2} f[x; y_1(x), y_2(x), \ldots, y_n(x); y_1'(x), y_2'(x), \ldots, y_n'(x)] \, dx. \tag{9.64}$$

Defining (as an extension of (9.4))

$$y_i(x, \epsilon) = y_i(x) + \epsilon\eta_i(x), \tag{9.65}$$

where $i = 1, 2, \ldots, n$, and where the $\eta_i(x)$ are n independent arbitrary functions satisfying

$$\eta_i(x_1) = \eta_i(x_2) = 0, \tag{9.66}$$

we obtain, in a similar way to the proof given in Section 9.2, the set of n Euler equations

$$\frac{\partial f}{\partial y_i} - \frac{d}{dx}\left(\frac{\partial f}{\partial y_i'}\right) = 0, \quad (i = 1, 2, \ldots, n). \tag{9.67}$$

All of these equations must be satisfied simultaneously for (9.64) to have a stationary value.

9.6 Lagrange's equations of dynamics

Consider a dynamical system made up of a given number of particles. Let $q_1, q_2, \ldots, q_n$ be n independent quantities (functions of time t) in terms of which it is possible to specify uniquely the position of each particle of the system. Such quantities are called the generalised coordinates of the system and may be, for example, distances or angles, and n is called the number of degrees of freedom of the system. For a single particle in three-dimensional space we need only

Figure 9.4

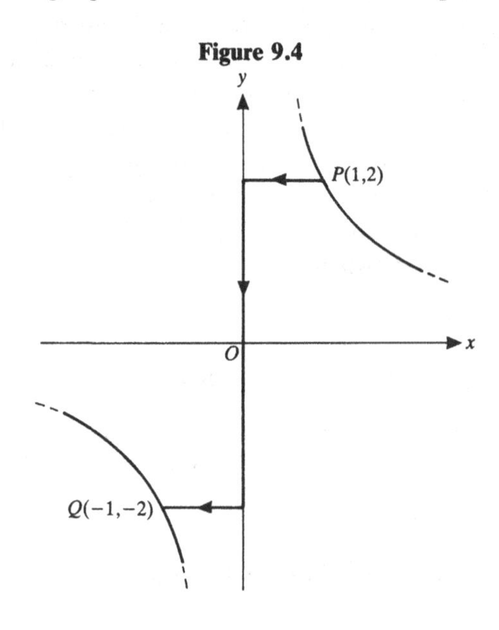

three coordinates to specify uniquely its position at time t. Clearly we could choose $q_1 = x$, $q_2 = y$, $q_3 = z$ or, in terms of spherical polar coordinates (see (1.84)), $q_1 = r$, $q_2 = \theta$, $q_3 = \phi$. Similarly, for two particles we require six coordinates. In the case of a rigid body, say a solid ellipsoid, six generalised coordinates are required: three to specify a point in the body, two to specify an axis through the point and a further one to represent the rotation of the body about this axis.

It can be shown that the Newtonian equations of motion for a system with kinetic energy T and potential energy V can be derived by requiring the quantity $\int_{t_1}^{t_2} (T - V)\, dt$ to be stationary, where T and V are, in general, functions of the $q_1, q_2, \ldots, q_n$ and their time derivatives. It is common to call $T - V = L$ the Lagrangian of the system. The condition for

$$\int_{t_1}^{t_2} (T - V)\, dt = \int_{t_1}^{t_2} L\, dt \tag{9.68}$$

to be stationary is (by Section 9.5) that L satisfies the set of n Euler equations

$$\frac{\partial L}{\partial q_i} - \frac{d}{dt}\left(\frac{\partial L}{\partial \dot{q}_i}\right) = 0, \tag{9.69}$$

where $L = L(t; q_1, q_2, \ldots, q_n; \dot{q}_1, \dot{q}_2, \ldots, \dot{q}_n)$ and $\dot{q}_i = dq_i/dt$. Equations (9.69) are known as Lagrange's equations and provide a powerful technique for deriving the equations of motion of a dynamical system. This is illustrated by the following two examples.

Example 7 We consider the equation of motion of a simple pendulum which was discussed in Chapter 3. A particle of mass m is suspended from a point O by a weightless inextensible string of length l (see Figure 9.5) and allowed to oscillate in the vertical plane. The position of the particle at time t is uniquely specified by the angle θ made by the string with the vertical and hence the system has one degree of freedom. We therefore choose θ as the generalised coordinate to describe the motion.

The kinetic energy T of the system is

$$T = \tfrac{1}{2}mv^2 = \tfrac{1}{2}m(l\dot{\theta})^2, \tag{9.70}$$

where $v = l\dot{\theta} = l\, d\theta/dt$ is the velocity of the particle. Likewise, the potential energy measured from the horizontal through O is

$$V = -mgl \cos\theta. \tag{9.71}$$

Hence

$$L = T - V = \tfrac{1}{2}ml^2\dot{\theta}^2 + mgl\cos\theta. \tag{9.72}$$

Since there is only one generalised coordinate, the single Lagrange equation (from (9.69)) is

$$\frac{\partial L}{\partial \theta} - \frac{d}{dt}\left(\frac{\partial L}{\partial \dot{\theta}}\right) = 0, \tag{9.73}$$

which, using (9.72), gives

$$-mgl\sin\theta - \frac{d}{dt}(ml^2\dot{\theta}) = 0 \tag{9.74}$$

or

$$\frac{d^2\theta}{dt^2} + \frac{g}{l}\sin\theta = 0. \tag{9.75}$$

This is the equation of motion discussed in Section 3.9 and has elliptic integral solutions. ◢

Example 8 Three masses m_1, m_2 and m_3 lie on a straight line and are connected by two springs of stiffness K for which the potential energy is (by Hooke's Law) $V = \tfrac{1}{2}K\,(\text{extension})^2$. No other forces act. If x_1, x_2 and x_3 are the displacements when the system is disturbed from its equilibrium position then

$$L = T - V = \tfrac{1}{2}m_1\dot{x}_1^2 + \tfrac{1}{2}m_2\dot{x}_2^2 + \tfrac{1}{2}m_3\dot{x}_3^2 - \tfrac{1}{2}K(x_2 - x_1)^2 - \tfrac{1}{2}K(x_3 - x_2)^2. \tag{9.76}$$

The Lagrange equations (9.69) give (with $q_i = x_i$)

$$\frac{\partial L}{\partial x_i} - \frac{d}{dt}\left(\frac{\partial L}{\partial \dot{x}_i}\right) = 0 \tag{9.77}$$

Figure 9.5

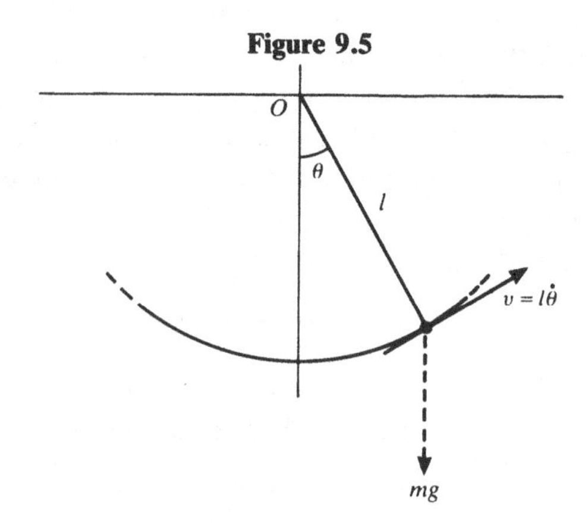

for $i = 1, 2$, or 3, whence

$$m_1\ddot{x}_1 - K(x_2 - x_1) = 0, \tag{9.78}$$

$$m_2\ddot{x}_2 + K(x_2 - x_1) - K(x_3 - x_2) = 0, \tag{9.79}$$

$$m_3\ddot{x}_3 + K(x_3 - x_2) = 0. \tag{9.80}$$

This coupled set of three second-order differential equations defines the motion of the system, and elementary methods may be used to obtain x_1, x_2 and x_3 as functions of time. ◢

9.7 Integrals involving derivatives higher than the first

Suppose we require that

$$I[y(x)] = \int_{x_1}^{x_2} f(x; y, y', y'', \ldots, y^{(n)})\, dx, \tag{9.81}$$

where $y^{(n)} = d^n y/dx^n$, should be stationary subject to the end conditions

$$y(x_1) = a_1,\ y'(x_1) = a_2, \ldots, y^{(n-1)}(x_1) = a_n, \tag{9.82}$$

$$y(x_2) = b_1,\ y'(x_2) = b_2, \ldots, y^{(n-1)}(x_2) = b_n. \tag{9.83}$$

Then by an analysis similar to that of Section 9.3 we find the generalised Euler equation (given here without proof)

$$\frac{\partial f}{\partial y} - \frac{d}{dx}\left(\frac{\partial f}{\partial y'}\right) + \frac{d^2}{dx^2}\left(\frac{\partial f}{\partial y''}\right) - \ldots + (-1)^n \frac{d^n}{dx^n}\left(\frac{\partial f}{\partial y^{(n)}}\right) = 0. \tag{9.84}$$

Example 9 Consider the integral

$$I[y(x)] = \int_0^{\pi} (y''^2 - 3y'^2 - 4y^2)\, dx, \tag{9.85}$$

given

$$y(0) = 0, \quad y'(0) = 0, \tag{9.86}$$

$$y(\pi) = 0, \quad y'(\pi) = 1. \tag{9.87}$$

The Euler equation (9.84) now gives

$$-8y - \frac{d}{dx}(-6y') + \frac{d^2}{dx^2}(2y'') = 0, \tag{9.88}$$

or

$$\frac{d^4y}{dx^4} + 3\frac{d^2y}{dx^2} - 4y = 0. \tag{9.89}$$

The solution of this equation is

$$y = Ae^x + Be^{-x} + C\cos(2x) + D\sin(2x), \tag{9.90}$$

where A, B, C and D are to be found by imposing the given conditions (9.86) and (9.87). Hence

$$A = -\frac{1}{2(1-e^{\pi})}, \quad B = \frac{1}{2(1-e^{-\pi})}, \quad C = \frac{\sinh\pi}{2(1-\cosh\pi)}, \quad D = \tfrac{1}{4}. \tag{9.91}$$

The stationary value of I is obtained by inserting (9.90) and (9.91) into (9.85). ◢

9.8 Problems with constraints

Many physical problems (especially within the area of control theory) require that an integral be maximised or minimised subject to constraints. Such problems are often called isoperimetric, the name given to the problem of finding the closed curve with a given fixed perimeter which bounds a maximum area. The solution of this problem is known to be a circle, a result which may be proved using the following analysis.

We require that

$$I[y(x)] = \int_{x_1}^{x_2} f(x, y, y')\,dx \tag{9.92}$$

(see (9.2)) be stationary subject to the integral constraint

$$J[y(x)] = \int_{x_1}^{x_2} g(x, y, y')\,dx = C, \tag{9.93}$$

where $g(x, y, y')$ is a given function and C is a given constant. As with the conventional Lagrange multiplier approach for finding the stationary points of a function subject to a constraint, we now form the linear combination

$$f + \lambda g, \tag{9.94}$$

where λ is a constant (Lagrange) multiplier, and determine the extremals of the integral

$$I + \lambda J = \int_{x_1}^{x_2} (f + \lambda g)\,dx. \tag{9.95}$$

The Euler equation appropriate to (9.95) is

$$\frac{\partial}{\partial y}(f+\lambda g)-\frac{d}{dx}\left(\frac{\partial(f+\lambda g)}{\partial y'}\right)=0, \tag{9.96}$$

which when solved leads to extremal curves of the general form

$$y=y(x,\lambda,C_1,C_2), \tag{9.97}$$

where C_1 and C_2 are constants. The three constants λ, C_1 and C_2 are uniquely determined by the three requirements that (9.97) passes through the end points x_1 and x_2, and that J has the prescribed value C as given by (9.93).

Example 10 To find the extremal curve $y=y(x)$ of the integral

$$I[y(x)]=\int_0^1 (y'^2+2yy')\,dx, \tag{9.98}$$

with $y(0)=0$ and $y(1)=0$, and where $y(x)$ is subject to the constraint

$$\int_0^1 y(x)\,dx=\tfrac{1}{6}, \tag{9.99}$$

and to determine the stationary value of $I[y(x)]$.

As in (9.95) we now consider the integral

$$\int_0^1 (y'^2+2yy'+\lambda y)\,dx. \tag{9.100}$$

Since the integrand of (9.100) is explicitly independent of x, we find, using (9.47), that the Euler equation is

$$y'^2+2yy'+\lambda y-y'\frac{\partial}{\partial y}(y'^2+2yy'+\lambda y)=\text{constant}. \tag{9.101}$$

Then

$$y'^2-\lambda y=C, \tag{9.102}$$

where C is a constant, so that

$$y'=\sqrt{(C+\lambda y)}. \tag{9.103}$$

Separating the variables and integrating, we have

$$y=\frac{\lambda}{4}(x+A)^2-\frac{C}{\lambda}, \tag{9.104}$$

where A is a constant.

Now $y = 0$ when $x = 0$. Hence

$$\frac{\lambda A^2}{4} = \frac{C}{\lambda}. \tag{9.105}$$

Also $y = 0$ when $x = 1$. Hence

$$\frac{\lambda}{4}(1 + A)^2 = \frac{C}{\lambda}. \tag{9.106}$$

Solving (9.105) and (9.106) gives

$$A = -\tfrac{1}{2}, \quad C = \lambda^2/16, \tag{9.107}$$

and consequently

$$y(x) = \frac{\lambda}{4}(x - \tfrac{1}{2})^2 - \frac{\lambda}{16} = \frac{\lambda}{4}(x^2 - x). \tag{9.108}$$

Now by (9.99) we require

$$\int_0^1 y(x)\, dx = \frac{\lambda}{4}\int_0^1 (x^2 - x)\, dx = \tfrac{1}{6}, \tag{9.109}$$

from which we find $\lambda = -4$. The solution for the extremal curve is therefore

$$y = x - x^2. \tag{9.110}$$

Inserting (9.110) into (9.98) we obtain the extremal value $I = \frac{1}{3}$. ◢

9.9 Direct methods and eigenvalue problems

The Euler equations obtained by requiring integrals to be stationary are, in general, non-linear ordinary differential equations. These frequently require numerical solution. An alternative and more direct approach to finding the minimum value of an integral was first developed by Rayleigh and Ritz. Suppose we again consider for simplicity the integral (9.2)

$$I[y(x)] = \int_{x_1}^{x_2} f(x, y, y')\, dx, \tag{9.111}$$

where $y(x_1) = \alpha$ and $y(x_2) = \beta$, α and β being fixed. Now consider a sequence of functions

$$y_1(x), y_2(x), \ldots, y_n(x), \ldots, \tag{9.112}$$

each satisfying the end conditions $y_n(x_1) = \alpha$, $y_n(x_2) = \beta$, for all n, such that

$$\lim_{n\to\infty} I[y_n(x)] = I_0 \tag{9.113}$$

and

$$\lim_{n\to\infty} y_n(x) = y_0(x), \tag{9.114}$$

where I_0 is the minimum value of I and $y_0(x)$ is the solution of the Euler equation producing the value I_0. The sequence (9.112) is called a minimum sequence. The simplest case is if $\alpha = \beta = 0$, and then the end conditions can be satisfied if each function y_n is written as

$$y_n(x) = C_1\phi_1(x) + C_2\phi_2(x) + \ldots + C_n\phi_n(x), \tag{9.115}$$

where $\phi_1(x), \ldots, \phi_n(x)$ are n independent functions each satisfying $\phi_i(x_1) = \phi_i(x_2) = 0$, and $C_1, \ldots, C_n$ are arbitrary constants. Inserting the nth member of the sequence (9.115) into (9.111) we find

$$I[y_n(x)] = \int_{x_1}^{x_2} f(x, y_n, y_n')\,dx \tag{9.116}$$

$$= F(C_1, C_2, \ldots, C_n). \tag{9.117}$$

By the usual methods of determining the maxima and minima, we now find the stationary value of $I[y_n(x)]$ by equating the n partial derivatives of F with respect to $C_1, \ldots, C_n$ to zero:

$$\partial F/\partial C_1 = \partial F/\partial C_2 = \ldots = \partial F/\partial C_n = 0. \tag{9.118}$$

We now have n equations for the n values $C_1, \ldots, C_n$, and the form of the extremal curve $y = y_n(x)$ can then be found from (9.115). Provided the sequence converges to the correct extremal curve $y = y_0(x)$ as $n \to \infty$, we can obtain an increasingly accurate approximation to the true stationary value I_0.

Example 11 Consider

$$I[y(x)] = \int_0^1 (y'^2 + xy)\,dx, \tag{9.119}$$

with $y(0) = 0,\ y(1) = 0$.

Suppose we choose $\phi_1(x) = x(1-x)$ since this satisfies the end conditions. Then the first member of a minimum sequence is

$$y_1(x) = C_1 x(1-x). \tag{9.120}$$

Hence

$$y_1'(x) = C_1 - 2C_1 x \tag{9.121}$$

and

$$I[y_1(x)] = \int_0^1 [C_1^2(1-2x)^2 + C_1x^2(1-x)]\,dx \tag{9.122}$$

$$= \tfrac{1}{3}C_1^2 + \tfrac{1}{12}C_1 = F(C_1). \tag{9.123}$$

Hence

$$\frac{\partial F}{\partial C_1} = \frac{dF}{dC_1} = 0 \tag{9.124}$$

gives

$$C_1 = -\tfrac{1}{8}, \tag{9.125}$$

and with this value, (9.123) gives

$$I[y_1(x)] = -\tfrac{1}{192}. \tag{9.126}$$

To find the next member of the sequence $y_2(x)$, we now choose $\phi_2(x) = x^2(1-x)$ which again satisfies the end conditions. Then

$$y_2(x) = C_1x(1-x) + C_2x^2(1-x) \tag{9.127}$$

and

$$y_2'(x) = C_1(1-2x) + C_2(2x-3x^2). \tag{9.128}$$

Calculating $I[y_2(x)]$, we find

$$I[y_2(x)] = \tfrac{1}{3}C_1^2 + \tfrac{2}{15}C_2^2 + \tfrac{1}{3}C_1C_2 + \tfrac{1}{12}C_1 + \tfrac{1}{20}C_2. \tag{9.129}$$

Hence

$$\frac{\partial F}{\partial C_1} = \tfrac{2}{3}C_1 + \tfrac{1}{3}C_2 + \tfrac{1}{12} = 0 \tag{9.130}$$

and

$$\frac{\partial F}{\partial C_2} = \tfrac{1}{3}C_1 + \tfrac{4}{15}C_2 + \tfrac{1}{20} = 0. \tag{9.131}$$

Solving these two equations for C_1 and C_2, we find

$$C_1 = C_2 = -\tfrac{1}{12}, \tag{9.132}$$

giving

$$y_2(x) = -\tfrac{1}{12}x(1-x) - \tfrac{1}{12}x^2(1-x) = \tfrac{1}{12}(x^3 - x). \tag{9.133}$$

From (9.129) and (9.132), the extremal value of I is

$$I[y_2(x)] = -\tfrac{1}{180}. \tag{9.134}$$

We see that the Euler equation for I in (9.119) is

$$x - \frac{d}{dx}(2y') = 0 \tag{9.135}$$

or

$$y'' = \tfrac{1}{2}x. \tag{9.136}$$

Integrating we have

$$y' = \tfrac{1}{4}x^2 + A \tag{9.137}$$

and

$$y = \tfrac{1}{12}x^3 + Ax + B. \tag{9.138}$$

Applying the end conditions gives $B = 0$, $A = -\frac{1}{12}$ and hence we find

$$y = \tfrac{1}{12}(x^3 - x). \tag{9.139}$$

This is precisely the function found in (9.133), and hence the value of I in (9.134) obtained by the Rayleigh–Ritz method is exact in this case. We emphasise that this depended on our choice of ϕ_1 and ϕ_2. For this simple example, the Euler equation gives the extremal curve very easily, but nevertheless this example illustrates the use of the Rayleigh–Ritz method when the value of the extremal curve is zero at both ends. ◢

If the end values of the curve are not both zero, then a linear combination of the form (9.115) can still be written but we must ensure that each $y_n(x)$ still satisfies the end conditions. This usually results in relationships between the arbitrary constants, thereby reducing the number of independent constants to less than n. The stationary value of $I[y_n]$ is then found by minimising with respect to the independent constants remaining. We give an example to illustrate this.

Example 12 Consider

$$I[y(x)] = \int_0^1 (y'^2 + xy^2)\, dx, \tag{9.140}$$

with $y(0) = 0$, $y(1) = 1$.

Suppose we choose $\phi_1(x) = x$ and $\phi_2(x) = x^2$. If we first put $y_1(x) = C_1\phi_1(x)$ so that $y_1(0) = 0$ then we must have $C_1 = 1$ in order that $y_1(1) = 1$. This choice leaves no free constants. Next if we put

$$y_2(x) = C_1x + C_2x^2, \tag{9.141}$$

which immediately satisfies $y_2(0) = 0$, then at $x = 1$, we have

$$y_2(1) = 1 = C_1 + C_2, \tag{9.142}$$

so that

$$C_2 = 1 - C_1. \tag{9.143}$$

Hence

$$y_2(x) = C_1 x + (1 - C_1)x^2, \tag{9.144}$$

which satisfies the end conditions for all C_1. Using (9.144), we find

$$I[y_2(x)] = \int_0^1 \{[C_1 + 2(1 - C_1)x]^2 + x[C_1 x + (1 - C_1)x^2]^2\}\, dx, \tag{9.145}$$

and integrating gives

$$I[y_2(x)] = \tfrac{7}{20}C_1^2 - \tfrac{3}{5}C_1 + \tfrac{3}{2} = F(C_1). \tag{9.146}$$

Hence

$$dF/dC_1 = \tfrac{7}{10}C_1 - \tfrac{3}{5} = 0, \tag{9.147}$$

giving

$$C_1 = \tfrac{6}{7} \tag{9.148}$$

and the extremal value

$$I[y_2(x)] = \tfrac{87}{70}. \tag{9.149}$$

We have therefore generated an approximation to the extremal value and an approximation to the extremal curve, which from (9.144) and (9.148) is

$$y_2(x) = \tfrac{6}{7}x + \tfrac{1}{7}x^2. \tag{9.150}$$

We note that the Euler equation corresponding to (9.140) is simply $y'' = xy$, which has a solution in terms of Airy functions (see Section 2.9). However, although the extremal curve can be found by applying the end conditions, the evaluation of (9.140) is difficult. ◢

We conclude this section with an example to illustrate the use of the method in evaluating eigenvalues.

Example 13 Consider the simple equation

$$d^2y/dx^2 + \lambda y = 0, \tag{9.151}$$

subject to the boundary conditions $y(0) = 0$, $y(1) = 0$.

The exact solution of (9.151) is

$$y = A\cos(\sqrt{\lambda}\, x) + B\sin(\sqrt{\lambda}\, x). \tag{9.152}$$

Imposing the boundary conditions, we find $A = 0$ and

$$\lambda = n^2\pi^2, \tag{9.153}$$

where $n = 1, 2, 3, \ldots$. These values of λ (an infinite discrete set) are called the eigenvalues of the equation. Now consider

$$I[y(x)] = \int_0^1 (y'^2 - \lambda y^2)\, dx, \tag{9.154}$$

which has (9.151) as its Euler equation. We choose some trial function passing through the end points, say

$$y_1(x) = C_1 x(1-x). \tag{9.155}$$

Then

$$I[y_1(x)] = C_1^2(\tfrac{1}{3} - \tfrac{1}{30}\lambda) = F(C_1). \tag{9.156}$$

Hence

$$dF/dC_1 = 0 = 2C_1(\tfrac{1}{3} - \tfrac{1}{30}\lambda), \tag{9.157}$$

which gives $\lambda = 10$ as an estimate for the lowest eigenvalue. This is to be compared with the exact value $\lambda = \pi^2 \approx 9.87$ obtained by putting $n = 1$ in (9.153).

In order to improve the approximation, we choose, say

$$y_2(x) = C_1 x(1-x) + C_2 x^2(1-x) \tag{9.158}$$

which also passes through the end points. We now find

$$I[y_2(x)] = \tfrac{1}{3}C_1^2 + \tfrac{1}{3}C_1C_2 + \tfrac{2}{15}C_2^2 - \lambda[\tfrac{1}{30}C_1^2 + \tfrac{1}{30}C_1C_2 + \tfrac{1}{105}C_2^2] = F(C_1, C_2). \tag{9.159}$$

Hence

$$\partial F/\partial C_1 = (\tfrac{2}{3} - \tfrac{1}{15}\lambda)C_1 + (\tfrac{1}{3} - \tfrac{1}{30}\lambda)C_2 = 0 \tag{9.160}$$

and

$$\partial F/\partial C_2 = (\tfrac{1}{3} - \tfrac{1}{30}\lambda)C_1 + (\tfrac{4}{15} - \tfrac{2}{105}\lambda)C_2 = 0. \tag{9.161}$$

Equations (9.160) and (9.161) are a pair of homogeneous linear equations for C_1 and C_2. For a non-trivial solution we require that the determinant of the coefficients is zero. This gives

$$\lambda^2 - 52\lambda + 420 = (\lambda - 42)(\lambda - 10) = 0 \tag{9.162}$$

and hence

$$\lambda = 10, \quad \lambda = 42. \tag{9.163}$$

The first of these is the value we found above corresponding to $n = 1$ (for which the exact value is $\pi^2 \approx 9.87$), while the second corresponds to $n = 2$ (and an exact value of $\lambda = 4\pi^2 \approx 39.48$). ◢

Problems 9

1. Find the extremal curves and the stationary values of the following integrals:

(i)
$$\int_P^Q [(dy/dx)^2 + 2yx - y^2]\,dx,$$

where $P = (0, 0)$ and $Q = (\pi/2, \pi/2)$,

(ii)
$$\int_P^Q [2y \sin x + (dy/dx)^2]\,dx,$$

where $P = (0, \pi)$ and $Q = (\pi, 0)$.

2. Find the extremal curve of the integral

$$I[y(x)] = \int_P^Q \left(\frac{dy}{dx}\right)^2 \left(1 + \frac{dy}{dx}\right)^2 dx,$$

where $P = (0, 0)$ and $Q = (1, 2)$.

3. Find the extremal curves of the integral

$$I[y(x)] = \int_P^Q \frac{y'^2}{1 + y^2}\,dx,$$

where $P = (0, 0)$ and $Q = (1, 2)$.

4. Find the function $y(x)$ which makes the following integral stationary:

$$I[y(x)] = \int_0^{\pi/2} (y'^2 + 2xyy')\,dx,$$

where $y(0) = 0$, $y(\pi/2) = 1$, and $y(x)$ is subject to the constraint

$$\int_0^{\pi/2} y\,dx = \pi/2 - 1.$$

5. Minimise the integral

$$\int_{t_0}^{t_1} x\sqrt{[1 + (dx/dt)^2]}\,dt,$$

where the function $x(t)$ is subject to the constraint

$$\int_{t_0}^{t_1} \sqrt{[1 + (dx/dt)^2]}\,dt = \text{constant}.$$

Show that $x(t) = A + B\cosh[(t - C)/B]$, where A, B and C are constants.

6. A curve $y = y(x)$ meets the x-axis at $x = \pm a$, and has length πa between these points. Show that the curve which encloses maximum area between itself and the x-axis is the semicircle $x^2 + y^2 = a^2$ $(y \geqslant 0)$.
7. The flight-path of an aircraft lies in the (x, y) plane, and the aircraft moves with a speed $v(y)$ (that is, a function of y only). If its path (assumed to be smooth) between two fixed points is such as to minimise the flight-time, show that the equation of the path is given by

$$x = \int \frac{v(y)\,dy}{\sqrt{[A - v^2(y)]}} + B,$$

where A and B are constants.
8. A particle moving in the (x, y) plane has a speed $u(y)$ depending only on its distance from the x-axis; its direction of motion makes an angle θ with the x-axis which can be controlled to give the minimum time of transit between two points. If $u(y) = u_0 \mathrm{e}^{-y/h}$, where u_0 and h are constants, and if the particle, starting at $x = 0$, $y = 0$, is required to reach $(\frac{1}{4}\pi h, h)$ in least time, prove that its initial and final directions must be $\theta_0 = \tan^{-1}(1 - \sqrt{2}/\mathrm{e})$ and $\theta_1 = \tan^{-1}(\mathrm{e}\sqrt{2} - 1)$.
9. Find a functional which has

$$y'' + y + x = 0$$

as its Euler equation. Hence obtain an approximate solution of the equation given $y(0) = y(1) = 0$ by using the trial functions

(i) $y = Cx(1 - x)$,
(ii) $y = C_1 x(1 - x) + C_2 x^2(1 - x)$,

and finding the values of C, C_1 and C_2.

ANSWERS TO PROBLEMS

Chapter 1

1. $a_{11}x_1^2 + a_{22}x_2^2 + a_{33}x_3^2 + 2(a_{13}x_1x_3 + a_{12}x_1x_2 + a_{23}x_2x_3)$.
2. (i) j, (ii) none, (iii) i and j, (iv) i on each side.
3. 3, 3, all three components zero, all three components zero.
4. $2\delta_{ij}$, 6.
5. $S_{ij} = \dfrac{1}{\beta}\left[T_{ij} - \dfrac{\alpha}{(3\alpha+\beta)}\delta_{ij}T_{kk}\right]$.

7. $c_{ij} = \delta_{ij}a_pa_p - a_ia_j$.
9. $\partial^2 f_k/\partial x_r\,\partial x_s = 2\delta_{rs}x_k + 2\delta_{kr}x_s + 2\delta_{ks}x_r$.
10. $\alpha = 4$, (i) $\beta = 7$, (ii) $\beta = 11$.

Chapter 2

1. (i) $\Gamma(5) = 24$, (ii) $\sqrt{(\pi/a)}$.
4. $2\pi/3\sqrt{3}$, using (2.25) and (2.27).
6. (i) $\frac{1}{2}\mathrm{B}(\frac{3}{4}, \frac{1}{2})$, (ii) $\pi/\sqrt{2}$, using (2.25) and (2.27).
9. $AJ_\nu(e^x) + BY_\nu(e^x)$.
10. $\dfrac{1}{\sqrt{x}}(A\cos x + B\sin x)$.

12. $3x^{\frac{2}{3}}J_2(x^{\frac{1}{3}}) + C$.
13. (i) 0, (ii) $-\frac{1}{3}$.

Chapter 3

1. $\sqrt{2}\,x$.
2. $4/(x-3)^2$.
3. $y = x\ln\left|\dfrac{x}{Cx+1}\right| + Ax$.

4. $y = \ln\left|\dfrac{\ln x}{x+C}\right|$.

5. $y = \dfrac{1}{2^{\frac{1}{3}}}\dfrac{\lambda Ai'(-x/2^{\frac{1}{3}}) + Bi'(-x/2^{\frac{1}{3}})}{\lambda Ai(-x/2^{\frac{1}{3}}) + Bi(-x/2^{\frac{1}{3}})}$,

   where λ is an arbitrary constant.

Chapter 4

1. $x + \frac{1}{12}x^4 + \frac{1}{504}x^7$.

2. $B(1-x)$, $Ax^{\frac{4}{3}}\left(1 - \dfrac{x}{21} + \dfrac{x^2}{315} - \dots\right)$.

3. $x + x^3 + \frac{4}{5}x^5$.
4. $y_1 = 2 + x^2$, $z_1 = 3x^2$,
   $y_2 = 2 + x^2 + x^3$, $z_2 = 3x^2 + \frac{3}{4}x^4 + \frac{3}{5}x^5$,
   $y_3 = 2 + x^2 + x^3 + \frac{3}{20}x^5 + \frac{1}{10}x^6$,
   $z_3 = 3x^2 + \frac{3}{4}x^4 + \frac{6}{5}x^5 + \frac{3}{28}x^7 + \frac{3}{40}x^8$.
5. $\cos t + \mu[\frac{3}{8}t \cos t - \frac{3}{16}\sin t - \frac{1}{32}\sin(3t)]$.
6. $e^{-x^3/6}[A\cos(2x) + B\sin(2x)]$.

7. $\dfrac{1}{x}e^{\pm Nx^3/3}$.

8. $x^{\frac{1}{2}}I_{1/(n+2)}\left(\dfrac{1}{n+2}x^{(n+2)/2}\right)$, $x^{\frac{1}{2}}K_{1/(n+2)}\left(\dfrac{1}{n+2}x^{(n+2)/2}\right)$.

Chapter 5

2. (i) $\frac{1}{2}\ln 2 + i(-\pi/4 + 2k\pi)$, where $k = 0, \pm1, \pm2, \dots$,
   (ii) $\pi/2 + k\pi + (i/2)\ln 3$, where $k = 0, \pm1, \pm2, \dots$.
3. $i(\pi/2 + k\pi)$, where $k = 0, \pm1, \pm2, \dots$.
4. $v = 2xy + 4y - 2x + C$, $f(z) = z^2 + (4 - 2i)z + C$.
5. g and h constant.
6. All parts $\frac{1}{3} + \frac{5}{3}i$; $z^2 + 1$ is analytic for all z.
7. (i) (*a*) 0, (*b*) $2\pi i$,
   (ii) $-2\pi i$, $\pi i/2$.
8. (i) $2\pi i \sinh 1$, (ii) $4\pi i e/3$.
9. (i) 2π, (ii) $2\pi i/9$, (iii) $2\pi i$.
10. (i) $1 + z^2 + \frac{1}{3}z^4 + \dots$ $(R = \infty)$,
    (ii) $\sin 1 + (\cos 1)z + (2\cos 1 - \sin 1)z^2/2 + \dots$ $(R = 1)$,
    (iii) $\ln 2 + \frac{1}{2}z + \frac{1}{8}z^2 + \dots$ $(R = \pi)$.

11. (i) $1 - \dfrac{1}{3!\,z^2} + \dfrac{1}{5!\,z^4} - \dots$ (essential singularity at $z = 0$),

    (ii) $1/(z-\pi) - 2(z-\pi) + \frac{2}{3}(z-\pi)^3 - \dots$ (simple pole at $z = \pi$),

    (iii) $e^{\pi/2}\left[\dfrac{1}{(z-\pi/2)^2} + \dfrac{1}{(z-\pi/2)} - \frac{1}{3}(z-\pi/2) + \dots\right]$ (double pole at $z = \pi/2$).

12. (i) $\frac{1}{z}+2+3z+4z^2+\ldots,$

(ii) $\frac{1}{z^3}+\frac{2}{z^4}+\frac{3}{z^5}+\frac{4}{z^6}+\ldots,$

(iii) $\frac{1}{(z-1)^2}-\frac{1}{(z-1)}+1-(z-1)+\ldots,$

(iv) $\frac{1}{(z-1)^3}-\frac{1}{(z-1)^4}+\frac{1}{(z-1)^5}-\frac{1}{(z-1)^6}+\ldots.$

13. (i) -1,
(ii) 1,
(iii) simple pole at $z=2$, residue $\frac{1}{8}$,
double pole at $z=1$, residue 2,
pole of order 3 at $z=0$, residue $-\frac{17}{8}$.

14. $2k\pi-\mathrm{i}\ln a$, where $k=0, \pm1, \pm2, \ldots.$

Chapter 6

1. (i) $\frac{1}{12}\pi$, (ii) $(4-2\sqrt{3})\pi$.
3. (i) $\frac{2}{3}\pi$, (ii) π, (iii) $\frac{1}{6}\pi$, (iv) $\frac{1}{2}\pi$.
6. $\pi/3\sqrt{2}$.
8. $I=0$.
9. $2\pi\mathrm{i}e^{\pi\mathrm{i}(\alpha-1)}$.
11. (i) $\frac{1}{2}+\frac{1}{4}\pi^2\operatorname{cosech}^2\pi+\frac{1}{4}\pi\coth\pi$,
(ii) $\frac{7}{720}\pi^4$.

Chapter 7

1. (i) $\frac{s^3}{[(s-a)^2+a^2][(s+a)^2+a^2]}$, (ii) $\frac{3}{4}\sqrt{(\pi/s^5)}$.
2. $\frac{2}{(s-1)[(s-1)^2+4]}$.
3. (i) $\frac{1}{3}[\cos x-\cos(2x)]$, (ii) $-1+e^x-xe^x+\frac{1}{2}x^2e^x$.
4. (i) $\frac{1}{2}$, (ii) $\frac{1}{2}$, (iii) $1/\sqrt{2}$.
6. $f(x)=\frac{1}{2}+1/\pi\sqrt{x}$.
7. $\sin x$.
9. (i) $6e^{-x}-7e^{-2x}+2e^{-3x}$,
(ii) $\frac{1}{4}H(x-2)\{1-\cos[2(x-2)]\}+\cos(2x)+\sin(2x)$,
(iii) $e^{-x}[1+e^aH(x-a)]$.
10. $x(t)=-1+te^t+e^t$, $y(t)=2+te^t-e^t$.

11. $y(x)=C\mathscr{L}^{-1}\{(1-s)^{\lambda}/s^{1+\lambda}\}$, where C is a constant; when $\lambda=2$, $y=C(1-2x+\frac{1}{2}x^2)$.
12. (i) $\frac{1}{9}e^{-t}+\frac{1}{3}te^{2t}-\frac{1}{9}e^{2t}$, (ii) $\frac{1}{16}e^{-t}(1-2t^2)+\frac{1}{16}e^{t}(2t-1)$.
13. $\frac{1}{3}x+\frac{8}{9}\sin(3x)$.
14. $\sqrt{(\pi/\lambda)}e^{-s^2/4\lambda}$, $\mathscr{F}\{\delta(x)\}=1$.
15. $\dfrac{4a(a^2-3s^2)}{(s^2+a^2)^3}$.
16. (i) $\dfrac{s}{s^2+4}, \dfrac{2}{s^2+4}$,

    (ii) $(\pi/2)\operatorname{sgn} s$, the cosine transform does not exist.
18. $\dfrac{1}{2\pi(1+x^2)}$.

Chapter 8

1. (i) $u=\cos\left[\sqrt{\left(\dfrac{1+x^2-y^2}{2}\right)}\right]$, (ii) $u=e^{xy/(x-y)}$,

   (iii) $u=-1/\ln(x^3y^2)$, (iv) $u=\dfrac{1}{y}\sinh(x^2+y)$.
2. $\phi=\ln x+\frac{1}{4}(x^2+y^2-1)$.
3. $u=e^{-2x}[f(x+iy)+g(x-iy)]$, f, g arbitrary functions.
4. $u=x^2e^y+2\ln x+x^2+y^2$.
5. $\phi=\frac{1}{3}(x^2+y^2)+(1/x)\cos(y/x)e^{-y^2/x^2}$.
7. $t=a^2\ln 2/\omega^2K$.
10. $u(r,t)=-\dfrac{2a^2}{\pi^2cr}\displaystyle\sum_{n=1}^{\infty}\frac{(-1)^n}{n^2}\sin\left(\frac{n\pi r}{a}\right)\sin\left(\frac{n\pi ct}{a}\right)$.
11. $u(x,t)=\dfrac{b}{c\sqrt{(\pi a)}}\displaystyle\int_0^t\frac{(1-e^{ac^2t})}{\sqrt{(t-u)}}e^{-x^2/4a(t-u)}\,du-\frac{be^{-cx}}{ac^2}(1-e^{ac^2t})$.
13. $u(x,y)=\dfrac{1}{2\sqrt{(\pi\alpha)}}\displaystyle\int_{-\infty}^{\infty}\frac{e^{-s^2/4\alpha}}{\cosh\sqrt{(1+s^2)}}\cosh[\sqrt{(1+s^2)}y]e^{isx}\,ds$.
14. $u(x,t)=-\dfrac{2\nu}{\theta}\,\partial\theta/\partial x$,

    where

    $$\theta(x,t)=\frac{u_0}{a\nu}\sum_{n=1}^{\infty}\left\{\cos\left(\frac{n\pi x}{a}\right)e^{-\nu n^2\pi^2t/a^2}\frac{[1-(-1)^ne^{-u_0a/2\nu}]}{\left(\dfrac{u_0^2}{4\nu^2}+\dfrac{n^2\pi^2}{a^2}\right)}\right\}.$$

Chapter 9

1. (i) $y = x$, $I = \frac{1}{2}\pi(1 + \frac{1}{12}\pi^2)$,
   (ii) $y = \pi - x - \sin x$, $I = \frac{5}{2}\pi$.
2. $y = 2x$.
3. $y = \sinh(x \sinh^{-1} 2)$.
4. $y = 1 - \cos x$.
9. (i) $C = \frac{5}{18}$, (ii) $C_1 = \frac{71}{369}$, $C_2 = \frac{7}{41}$.

Index

Printed in the United States
By Bookmasters